P9-DTB-442

Ecological Stability

Ecological Stability

Edited by

M. B. USHER

and

M. H. WILLIAMSON

Department of Biology, University of York
York, England

The chapters of this book are based on, but are not identical to, twelve papers given to a 'workshop' meeting at York University on 17–18 July 1973.

LONDON
CHAPMAN AND HALL

First published 1974
by Chapman and Hall Ltd
11 New Fetter Lane, London EC4P 4EE

Printed in Great Britain by
T. & A. Constable Ltd., Edinburgh

ISBN 0 412 12610 9

Distributed in the U.S.A.
by Halsted Press, a Division of
John Wiley & Sons, Inc., New York

Contents

PART THREE TEMPORAL STUDIES BETWEEN TROPHIC LEVELS: PREDATOR-PREY SYSTEMS

PART FOUR SPATIAL STUDIES BETWEEN TROPHIC LEVELS

Preface

In the Autumn of 1971, Gordon Conway came up to York from Imperial College to give a talk about his research. Afterwards, talking about the present state of mathematical ecology in Great Britain, it seemed desirable to attempt to set up a discussion group of those working actively in mathematical ecology. As a result, a first meeting was held in York in July 1972, and by that time it was called a 'Workshop' on Mathematical Ecology. The meeting lasted two days and was attended by about 30 people, many of them from the Department of Zoology at Imperial College, and from the Department of Biology at York, but with a very satisfactory mixture of people from Fisheries Laboratories, Grassland Research, microbial work, and so on. At the end of the meeting the Workshop agreed on certain things: that we should attempt to continue with this Workshop annually for some years; that we should not affiliate with any of the national societies; and that we should continue to restrict attendance to those who had been invited. We wanted to keep the meeting small and informal, with not more than about 30 people, and we wanted to ensure that all those present were working actively in the field, wishing to exchange opinions with others in the group, and not merely there to listen or learn.

The major change we planned for 1973 was to concentrate papers on one particular topic, and after some discussion we chose 'Ecological Stability'. The 1973 meeting was also held in York, and was so successful that we thought it was worthwhile to publish papers based on the papers delivered and discussed at the Workshop. The result is this book. As we only decided to publish after the meeting, it has not been possible to include any of the discussions, but the papers as they appear here have been modified as a result of the discussions. It is not our intention to publish the papers after every Workshop, or even after any other Workshop apart from this second one. We thought it important to publish quickly, and so have used the camera-ready style. We hope to be able to see this volume published before the 1974 Workshop, which will be held at Imperial College on the topic of 'Heterogeneity'.

One of the main aims of the Workshop is to bring people together who are working on mathematical problems in ecology and who use quite different sorts of organisms and mathematical techniques. The diversity of approaches can be seen in the meanings that can be attached to the phrase 'ecological stability', which will become apparent in the various papers of the book. Roughly speaking ecological stability is the strength of the tendency for a population or set of populations to come to an equilibrium point or to a limit cycle, and also, related to that, the ability of a population system to counteract perturbations. The measurement of stability is a different problem from its definition, since it will not necessarily be the same in analytical models, computer simulations and in real populations. Amongst all these complications there are two aspects of the problem that are important: the trophic structure of the system, and the variability of the populations in time or space or both.

We have grouped the papers into four sections, corresponding to different parts of the spectrum of problems in ecological stability. Part I contains two papers that deal with stability in a single trophic level. The next two parts, which are the two largest parts of the book, deal with situations involving the interaction of one trophic level with another. The relationships that have been considered between the levels may be either predator-prey or host-parasite, a division that is somewhat artificial, but it will be seen to reflect the variety of topics discussed in the papers. Host-parasite systems are discussed in Part II, and include both host-helminth systems (which are discussed in the first three papers) and insect parasitoids and their hosts (in the fourth paper). Predator-prey systems are discussed in Part III. Part IV contains the only paper which deals not only with the interactions of the trophic levels, but also with various scales of spatial effects.

The success of the two Workshops held so far owes a lot to a number of people. In 1972 Brian McArdle, and in 1973 David Thompson gave much help to Mark Williamson, and in both years Margaret Britton attended to almost all the secretarial work. In the preparation of this book, we are greatly indebted to Ann Fisher for drawing many of the figures, and to Eileen Kelly for typing the book, and particularly for coping with the intricacies of mathematical typing.

York
January 1974

M.B.U.
M.H.W.

List of 'workshop' participants

The following list identifies all participants of the 'workshop' meeting held at York University, 17-18 July 1973. The list is compiled primarily in alphabetical order of government department/university and, where relevant, also in alphabetical order of participants. Those participants shown in upright typeface are contributors to this volume.

J.H. Steele	Department of Agriculture and Fisheries for Scotland, Marine Laboratory, P.O. Box 101, Victoria Road, Aberdeen AB9 8DB, U.K.
J. Horwood T. Wyatt	Ministry of Agriculture, Fisheries and Food, Fisheries Laboratory, Lowestoft, Suffolk, U.K.
G. Gettinby	New University of Ulster, Department of Mathematics, Coleraine, Co.Londonderry, U.K.
R.M. May	Princeton University, Department of Biology, Princeton, New Jersey, 08540, U.S.A.

G.R. Conway *M.J. Crawley* *P. Furniss* *M.P. Hassell* *G. Murdie* *G. Norton* *Lesley Rushton* *Christine Shoemaker*	University of London, Imperial College Field Station, Silwood Park, Sunninghill, Ascot, Berks, U.K.
R.M. Anderson	University of London, King's College, Department of Zoology, Strand, London WC2R 2LS, U.K.
M. Bazin Vija Rapa P.T. Saunders	University of London, Queen Elizabeth College, Atkins Building, Campden Hill Road, London W8 7AH, U.K.
M.I. Webber	University of Oxford, Edward Grey Institute of Field Ornithology, Department of Zoology, South Parks Road, Oxford OX1 3PS, U.K.
J.E. Cohen	University of Oxford, Genetics Laboratory, Department of Biochemistry, South Parks Road, Oxford OX1 3QU, U.K.
D.J. Bradley	University of Oxford, Harkness Laboratory, Department of Pathology, Radcliffe Infirmary, Oxford OX2 6HE, U.K.
S. Hubbard D.J. Rogers	University of Oxford, Hope Department of Entomology, University Museum, Oxford OX1 3PW, U.K.
R. Mead	University of Reading, Department of Applied Statistics, Whiteknights, Reading RG6 2AN, U.K.
R.H. Smith	University of Reading, Department of Zoology, Whiteknights, Reading RG6 2AN, U.K.
.R.W. Harris	University of Surrey, Department of Biological Sciences, Guildford, Surrey GU2 5XH, U.K.

J. Beddington
R. Bonser
J. Byrne
J. Craigon
M. Joyce
J.H. Lawton
B.C. Longstaff
B.M. McArdle
N. Nice
G. Smith
M. Speight
M. Storey
D. Thompson
M.B. Usher
M.H. Williamson
R.A. Wilson

University of York, Department of Biology, York YO1 5DD, U.K.

General introduction

ROBERT M. MAY

Biology Department, Princeton University

The ecology workshop held at York in 1973 aimed to provide a forum for informal discussion around the general theme of ecological stability. This book derives from the workshop; as such it contributes to the theoretical framework which is increasingly being seen to underlie the body of observation and description of plant and animal communities.

This is not to say that theoretical ecology aims for the exact relations and crisp determinacy that characterize the physical sciences. Rather it aims to perceive patterns and attain qualitative understandings. It could be said that the typical sign in the equations of the physical sciences is the 'equals' sign, $A = B$, whereas in population biology it is the 'order of magnitude' sign, $A \sim B$ (and in some of the social sciences the 'definition' sign, $A \equiv B$).

This stream of theory has many tributaries. One (associated with Odum, Margalef, and others) draws inspiration and analogies from thermo-dynamics, and is concerned with broad patterns of energy flow through food webs. A second theme, which reaches back to the origins of the subject, deals with the physical environment, and the way it limits species' distributions and affects community organization. A third tributary concentrates on the way biotic interactions between and within populations act as forces moulding community structure.

All these approaches are represented in this volume, although the third predominates. More specifically, most of the papers treat discrete generation predator-prey or host-parasite relations; that is, they focus upon various aspects

---

Ecological Stability
edited by M.B. Usher and M.H. Williamson.

of difference equations for biological interactions *between* trophic levels, and explore their implications for population dynamics. None of the papers deals with a situation in which competition is pre-eminent.

The preponderance of such papers reflects, at least in part, a difference in research emphasis in population biology between the U.K. and North America. An analogous conference in the U.S.A. would similarly have been dominated by studies of biological interactions between species, but they would, in the main, have been studies of differential equations. Moreover, much more attention would have been paid to competition; that is, to interactions *within* a given trophic level. This stems from historical patterns. Following early work, and particularly in the wake of Hutchinson and MacArthur, theoretical ecology in North America has tended to focus (often abstractly) on higher vertebrates, and consequently on overlapping generations and differential equations. (Witness the fact that of the six compendious North American textbooks reviewed by Orians (1973), only two even mention the Nicholson-Bailey equation, and that in passing.) In Britain, much of the best ecological theory has been motivated by practical concern with temperate insects, hence non-overlapping generations and difference equations. All this is, of course, a gross oversimplification, but I think it conveys a useful kernel of truth beneath the distracting husk of individual counter-examples.

There is much scope for cross-fertilization between the two schools. Thus broad themes of time-delay in feedback mechanisms, and of pervasive and natural stable limit cycles, developed for rather abstractly general differential equation models (May, 1973), can provide analytic insights into, and syntheses between, various difference equation models generated by concrete entomological and parasitological problems, some examples of which are to be found in this volume. This point has recently been developed in detail by May *et al.* (in press). Conversely, relatively detailed and realistic studies of the dynamics of insect predator-prey systems, and particularly of systems with patchy prey distributions and differential aggregation of the predators in response to such heterogeneities in prey density (as discussed, e.g., by Hassell and May (1973, in press), and touched upon obliquely here by Rogers, by Lawton, and by Harris) provide detailed insights into how spatially heterogeneous predator-prey systems can be a robust source of species diversity and ecological stability, thus adding much greater depth to the abstract models of Levins and Culver (1971) and Horn and MacArthur (1972).

In this general context, it may usefully be remarked that the third general class of interaction between species, namely mutualism, rarely receives systematic analytic treatment on either continent. All contemporary textbooks give extensive coverage of models for predator-prey and competition, but analogous models for mutualistic interactions are absent (although they may easily be constructed and studied). A part of the explanation may be that mutualism is less crucial in the dynamical workings of many communities. However, as pointed out by many people, there can be no doubt that the mutualistic interactions between plants and pollinators or seed-dispersers are central to an understanding of many plant communities, and full comprehension of ecological stability in plant systems will need to draw on all three strands, namely plant-herbivore (prey-predator), competition, and mutualism.

Another area where British ecology draws on a richer tradition is in the incorporation of explicit age structure into population models with discrete generations. Longstanding efforts towards extending the density independent Leslie matrix formalism towards density dependent difference equations (e.g., Leslie, 1948; Pennycuick *et al.*, 1968; Usher, 1972) are represented in this volume by Williamson's paper; by way of contrast, it is only recently that the general dynamical implications of age structure in differential equation models has received systematic attention (Oster and Takahashi, in press; Auslander, Oster and Huffaker, in press).

In attempting to relate any of the above theoretical considerations to a given body of field observations as to patterns of species relative abundance, it is first necessary to decide whether such observed patterns really do reflect aspects of community structure which devolve directly upon specific interactions between populations, or whether the patterns stem from the interplay of many independent factors, so that they reflect merely the statisticians' Central Limit Theorem. Some touchstone for resolving this question would clearly be useful.

I have now manoeuvered this rambling introduction to a point where it is natural to refer to my own contribution to the workshop. This contribution treated patterns of species relative abundance and diversity, and will be presented in full detail (May, in press) in a volume of papers dedicated to Robert MacArthur. It aims to be a synthesis and review of the major species abundance distributions which have been propounded in the literature, of their underlying statistical or biological origins, and of the contrasts between them as revealed by various indices of diversity or equitability.

In the remainder of this introduction, I shall first summarize my conclusions as to how the various distributions arise, secondly sketch explanations for the enigmatic and almost mystical rules $a \simeq 0{\cdot}2$ and Preston's canonical hypothesis (and the consequent species-area relation) for the lognormal distribution, and thirdly summarize the predictions the various distributions make for an assortment of indices of diversity and equitability. This third part contains material which will not be published elsewhere, and suggests a way of discriminating between biologically revealing, as against merely statistical, patterns.

PATTERNS OF SPECIES RELATIVE ABUNDANCE

If the community under study comprises a large number of species, fulfilling diverse roles, the pattern of relative abundance of species will be governed by the interplay of many more-or-less independent factors. As pointed out by MacArthur (1960), Whittaker (1972) and others, it is in the nature of the equations of population dynamics that these several factors should compound multiplicatively, and the Central Limit Theorem applied to such a product of factors implies a lognormal distribution. Hence a lognormal species abundance distribution is to be expected for any large and heterogeneous assembly of species; this distribution reflects the statistical law of large numbers, and says nothing about the underlying biology.

If attention is restricted to communities with a limited number of taxonomically similar species, in stable competitive contact with each other in a relatively homogeneous habitat, the biological interactions are likely to dictate a species abundance distribution significantly *more uniform* than the lognormal. As pointed out by MacArthur (1957, 1960), and thoroughly discussed by Webb (in press), if the underlying biological picture is one of intrinsically uniform division of some major environmental resource, the statistical outcome is the well-known 'broken stick' distribution; this is the distribution relevant, for example, to collecting plastic Wee Beasties out of weeties packets, assuming the various Beastie species have a uniform distribution at the factory. Although observation of a broken stick distribution does not validate the very specific model initially propounded by MacArthur, it does indicate that some major factor is being roughly evenly apportioned among the community's constituent species, and thus that biological (rather than merely statistical) information may be extracted from the data.

Conversely, distributions of relative abundance significantly *less uniform* than lognormal can arise in relatively small and simple communities dominated (as are some early successional communities) by extreme 'niche pre-emption'. In the ideal case, the most successful or first species pre-empts a fraction k of some governing resource, the next species a fraction k of the remaining niche space, and so on, to give a geometric series distribution of relative abundance (characterized by the parameter k). A statistically more realistic expression of these ideas commonly leads to a logseries distribution of relative abundance (conventionally characterized by the parameter α, which is essentially $1/k$). Observation of such highly non-uniform distributions again tells us something about the underlying biology.

These bald statements are substantiated and developed more fully elsewhere (May, in press). It is important to note that the remarks pertain to the actual distribution of species relative abundance, *per se*, and pay no attention to sampling problems which may in practice becloud the issue. That is, we deal with the distribution of species abundance as known to God, as it were, and beg all questions of human frailty in attaining His knowledge. This is not to say sampling problems are uninteresting, but rather that they constitute a separate area of study.

The above themes have illuminating parallels elsewhere. Thus the distribution of wealth in the U.S.A. could be expected to arise from the random interplay of many independent factors, and hence to be lognormal; this is indeed the fact. The distribution of wealth in the U.K., particularly at the upper end, is significantly less uniform than lognormal, being described better by a logseries; this has been interpreted as arising from the mechanisms whereby the bulk of inherited wealth passes to the eldest son, an example of 'niche pre-emption'. Conversely, the distribution of income (rather than wealth) in Scandanavian countries is significantly more uniform than lognormal, reflecting a predominance of deliberate regulatory mechanisms over statistical chaos.

To reiterate, if the pattern of relative abundance arises from the interplay of many independent factors, as it must once the number of species is large, a lognormal distribution is both predicted by theory and usually found in nature. In relatively small and homogeneous sets of species, where a single factor can predominate, one limiting case (which may be idealized as a perfectly uniform distribution) leads to MacArthur's broken stick distribution, while the opposite limit (which may be idealized as a geometric series) leads to a logseries distribution. These two extremes correspond to patterns of relative abundance which are, respectively, significantly more even, and significantly less even, than the lognormal pattern. That is, the lognormal reflects the statistical Central Limit Theorem; conversely, in those special circumstances where broken stick, geometric series or logseries distributions are observed, they reflect features of the community biology.

The obvious way to discover which distribution best describes a given collection of data is to study the relative abundance in full detail. It is, however, useful to contemplate some single index which may characterize the distribution and suggest which pattern it tends to fit. We address this problem below, after first pausing to make some further remarks about lognormal distributions.

TWO ASPECTS OF LOGNORMAL DISTRIBUTIONS

There are two empirical 'laws' relating to lognormal species abundance distributions, the significance of $a \simeq 0{\cdot}2$ and Preston's canonical hypothesis, both of which have provoked considerable speculation in the ecological literature. To enunciate these laws, we must first display the lognormal distribution of species abundance in its conventional form (Preston, 1948, 1962; Whittaker, 1970, 1972)

$$S(R) = S_0 \exp(-a^2R^2) \qquad (1$$

Here $S(R)$ is the number of species in the Rth 'octave', and R is the logarithm (to the base 2) of the species abundance N, referred to the modal species' population, N_0; so $R = \log_2 (N/N_0)$. S_0 is the maximum or modal value of $S(R)$, attained at $N = N_0$, and a is an inverse measure of the width of the distribution. From Equation 1 it follows that the distribution of the total number of animals (i.e., number of species times number of individuals per species) is also lognormal, with the peak of this distribution displaced to the right of the peak in the $S(R)$ distribution.

The octave, R_N, in which the distribution of total individuals peaks can be related to the octave where we expect to find the single most abundant species, R_{max}, by a parameter γ, which is related to the familiar parameters a and S_0 of Equation 1 by

$$\gamma \equiv \frac{R_N}{R_{max}} = \frac{\ln 2}{2a(\ln S_0)^{0\cdot5}} \tag{2}$$

For a fuller discussion, see MacArthur and Wilson (1967) or May (in press).

The first empirical law (originally noted by Hutchinson in 1953) is that, for collections of data ranging from diatoms to moths to birds,

$$a \simeq 0\cdot2 \tag{3}$$

It is perhaps already disquieting to notice that other lognormal distributions, such as that of wealth in the U.S.A., or of human populations among nations of the world, or of G.N.P. of nations, also have $a \simeq 0\cdot2$.

The second observation is Preston's 'canonical hypothesis' (Preston, 1948, 1962),

$$\gamma = 1 \tag{4}$$

This rule fits a diversity of data, and is the basis of Preston's (1962) and MacArthur and Wilson's (1967) explanation of many species-area relations.

Both these rules are tantalizingly general. In view of the speculation they have prompted, it is surprising that no theoretical explanation has previously been attempted. Such an explanation is now outlined.

First, assume the lognormal distribution to be canonical, $\gamma = 1$. Equation 1 now provides a unique relation between the parameter a and the total number of species in the community, S_T. This relation is studied in May (in press), and it is seen that a depends very weakly on the actual value of S_T once there are more than ten or so species; as S_T ranges from 20 to 10,000 species, a varies from 0•30 to 0•13. Indeed, the dependence of a on S_T scales as $(\ln S_T)^{-0\cdot5}$ once $S_T \gg 1$, which is a very weak dependence. In short, the rule $a \simeq 0\cdot2$ is a mathematical property of the canonical lognormal.

It remains to consider the canonical hypothesis itself. Without this hypothesis, the general lognormal is characterized by two parameters, commonly a and S_0 (Equation 1), but equivalently a and γ. These two parameters can be uniquely related to the total number of species, S_T, and the total number of individuals, J (expressed in units of the population of the rarest species). The form of this relation is illustrated and discussed in detail in May (in press), where it is shown that communities with S_T ranging from around 20 to 10,000, and with J ranging from around $10S_T$ to $10^7 S_T$, are characterized by values of a in the range 0•1 to 0•4, and of γ from 0•5 to 1•8. Thus this enormous range of communities is roughly consistent with the rules $a \simeq 0{\cdot}2$ and $\gamma \simeq 1$. Again, the basic reason is that for $S_T >> 1$, the quantities a and γ depend on S_T and J only as $(\ln S_T)^{-0\cdot 5}$ and $(\ln J)^{-0\cdot 5}$.

This admittedly is only an imprecise and qualitative explanation of the rules expressed by Equations 3 and 4. However, these empirical rules are themselves only rough ones, and a qualitative theory seems appropriate. The rules are seen as mathematical properties of the lognormal distribution, not reflecting anything biological.

The moral surely is that one should seek to characterize the statistics of the lognormal distribution by indices more sensitive than a and γ.

The most remarkable success of Preston's canonical hypothesis is in its island biogeographical application to species-area studies (Preston, 1962; MacArthur and Wilson, 1967). Assuming $\gamma = 1$ leads, via Equations 1 and 4, to a unique relation between the total number of species, S_T, and the total number of individuals, J. In conjunction with the biological assumption that J is proportional to th area, A, wherein the community is contained (as discussed by MacArthur and Wilson, and by Preston), this leads to a unique relation between S_T and A. For large S_T, the relation has the approximate form

$$\ln S_T = z \ln A + (\text{constant}) \quad (5$$

with $z = 0{\cdot}25$; for smaller S_T, the dependence on A tends to become steeper. This relation accords with a large and growing body of species-area data on communities of species isolated on real or virtual islands; the 14 studies reviewed by May (in press) conform to Equation 5 with z ranging from 0·21 to 0·37, with an average value of 0·28. If one abandons Preston's empirical canonical hypothesis, then a general lognormal distribution is still to be expected for any large and heterogeneous collection of species, and moreover it can be shown that Equation 5 still applies, with $z = (1+\gamma)^{-2}$ if $\gamma < 1$ and $z = 1/4\gamma$ if $\gamma > 1$. Thus a range of γ values from 0·6 to 1·4 corresponds to a spread of z values from 0·39 to 0·18, indicating that MacArthur and Wilson's successes are not narrowly specific to the canonical lognormal, but are in effect yet another robust manifestation of the Central Limit Theorem.

EQUITABILITY INDICES

Various indices have been proposed as measures of the 'diversity' of a community of species. Defining p_i as the number of individuals in the ith species divided by the total number of individuals, the Shannon-Weaver diversity index is

$$H \equiv - \sum_i p_i \ln p_i \tag{6}$$

Alternatively, the Simpson (1949) diversity index D is defined as

$$D \equiv \frac{1}{\sum_i p_i^2} \tag{7}$$

A dominance index d may be defined as the fraction of individuals in the most abundant species,

$$d \equiv (p_i)_{max} \tag{8}$$

A variety of other such indices of diversity and dominance may similarly be constructed.

Then, given any of the species abundance distributions discussed above, it is a routine statistical exercise to calculate the expected value of any of these indices, as functions of S_T and other relevant parameters. Although less conventional, it is illuminating also to calculate the standard deviations about the mean value of these indices, for the various distributions. A detailed discussion, and comparison between the different distributions, is in May (in press). The tendencies shown by these results are revealing, and corroborate the remarks made earlier as to the essential character of the distributions. The broken stick (and the ideal uniform distribution) is notably more diverse (larger H, larger D, smaller d) than the logseries (and the ideal geometric series), with the lognormal occupying an intermediate position.

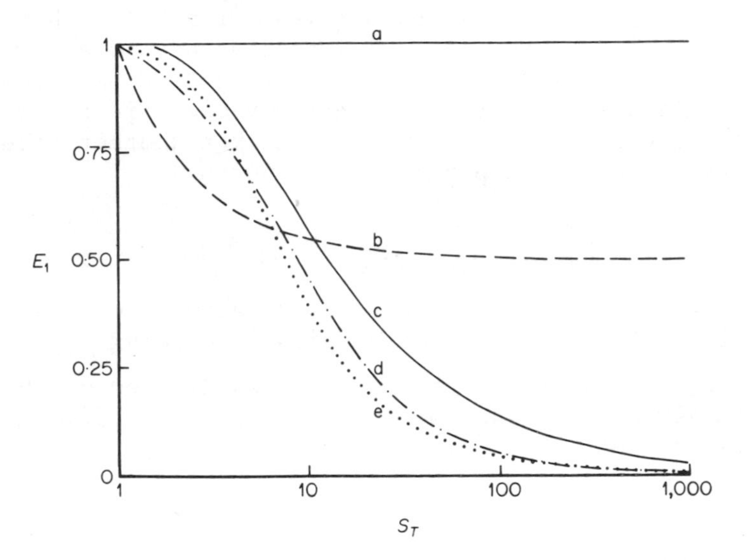

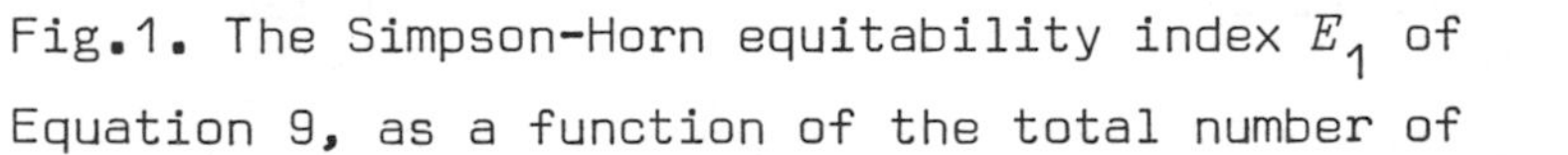

Fig.1. The Simpson-Horn equitability index E_1 of Equation 9, as a function of the total number of species in the community, S_T (plotted on a logarithmic scale), for various species relative abundance distributions. The curve labelled (a) is the completely uniform distribtuion; (b) is the broken stick; (c) is the canonical lognormal; (d) is a geometric series distribution (with k = 0•4); and (e) is a logseries distribution (with α = 5)

An alternative way to express these results is in terms of 'equitability' indices, which tend to unity in the limit of a community with many equally abundant species, and to zero in the opposite limit of a highly non-uniform distribution of relative abundance. Thus 'equitability' is measured on a 0 to 1 scale. A comparative anatomy of such equitability indices is of some interest, and has not been published elsewhere.

One index, which may be called the *Simpson-Horn equitability index* (see, e.g., Horn, 1971) is

$$E_1 \equiv D/S_T \tag{9}$$

Alternatively, noting that H itself is a weak measure of diversity, one may define the *MacArthur-Terborgh index* (MacArthur, 1965; Terborgh, in press) as

$$E_2 \equiv e^H/S_T \tag{10}$$

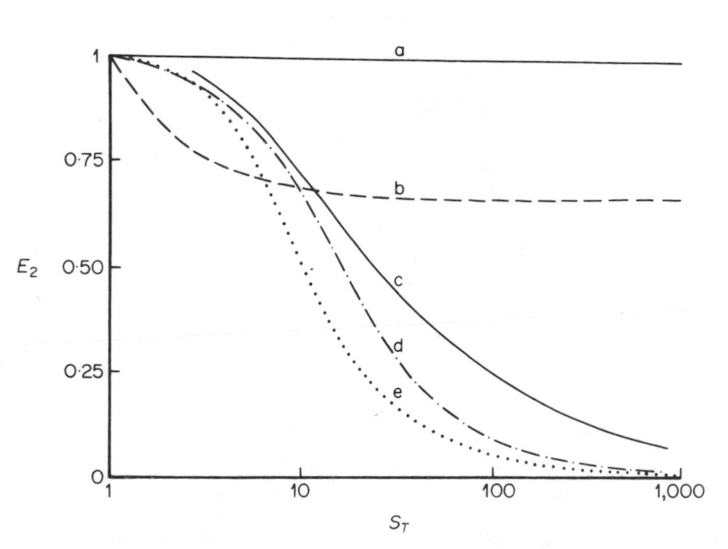

Fig.2. The MacArthur-Terborgh equitability index E_2 of Equation 10 as a function of S_T. The labelling is exactly as in Fig.1

These indices are illustrated for the various species relative abundance distributions, as functions of the total number of species S_T, in Figs.1 and 2 respectively. The asymptotic results for large S_T are summarized in Table 1.

It can be seen that once the total number of species in the community, S_T, is moderately large, these equitability indices are transparent measures of the character of the distributions. The relatively uniform broken stick distribution settles to a constant value; the highly non-uniform niche pre-emption distributions scale as $1/S_T$; while the lognormal displays intermediate behaviour.

An example of the practical use of these indices is provided by Terborgh's studies of equilibrium (and 'super equilibrium') bird communities in the neotropics. He finds his data to be characterized by values of E_2 in the vicinity

Table 1. Asymptotic relations between the equitability indices E_1, E_2 and S_T, for $S_T \gg 1$

Distribution	E_1 see Equation 9	E_2 see Equation 10
Canonical lognormal	$\frac{(4\cdot53)\ln S_T}{S_T}$	$\frac{\exp(1\cdot13\lvert\ln S_T\rvert^{0\cdot5})}{S_T}$
Uniform	1·000	1·000
Broken stick	0·500	0·655
Geometric series (parameter k)	$\frac{2-k}{k}\frac{1}{S_T}$	$\frac{(1-k)^{1-1/k}}{k\,S_T}$
Logseries (parameter α)	$\frac{\alpha}{S_T}$	$\frac{(1\cdot78)\,\alpha}{S_T}$

of 0·7–0·8, for communities comprising some 50 species (for $S_T = 50$, the canonical lognormal would suggest $E_2 = 0\cdot35$). He can thus be reasonably certain that the community patterns revealed by his studies are determined by biological interactions, and not merely by the Central Limit Theorem.

ACKNOWLEDGEMENTS

The success of the workshop was mainly due to the organizational efforts by people at the University of York, particularly Mark Williamson; it is a pleasure to acknowledge them. I also wish to thank G.R. Conway, M.P. Hassell and T.R.E. Southwood, who were responsible for my presence in England (under the auspices of their Ford Foundation Grant) at the time of the workshop, and who contributed to some of the ideas developed here. Finally I thank J. Beddington, H.S. Horn, Donna Howell and J.W. Terborgh whose critical comments and suggestions have shaped this manuscript.

REFERENCES

Auslander, D., Oster, G. and Huffaker, C. (in press).

Dynamics of interacting populations.
Hassell, M.P. and May, R.M. (1973). Stability in insect host-parasite models. *J. Anim. Ecol.*, 42, 693-726.
Hassell, M.P. and May, R.M. (in press). Aggregation of predators and insect parasites and its effect on stability.
Horn, H.S. (1971). *The Adoptive Geometry of Trees*. Princeton: Princeton University Press.
Horn, H.S. and MacArthur, R.H. (1972). Competition among fugitive species in a harlequin environment. *Ecology*, 53, 749-52.
Hutchinson, G.E. (1953). The concept of pattern in ecology. *Proc. Acad. nat. Sci. Philad.*, 105, 1-12.
Leslie, P.H. (1948). Some further notes on the use of matrices in population mathematics. *Biometrica*, 35, 213-45.
Levins, R. and Culver, D. (1971). Regional coexistence of species and competition between rare species. *Proc. natn. Acad. Sci. U.S.A.*, 68, 1246-8.
MacArthur, R.H. (1957). On the relative abundance of bird species. *Proc. natn. Acad. Sci. U.S.A.*, 43, 293-5.
MacArthur, R.H. (1960). On the relative abundance of species. *Am. Nat.*, 94, 25-36.
MacArthur, R.H. (1965). Patterns of species diversity. *Biol. Rev.*, 40, 510-33.
MacArthur, R.H. and Wilson, E.O.(1967). *The Theory of Island Biogeography*. Princeton: Princeton University Press.
May, R.M. (1973). *Stability and Complexity in Model Ecosystems*. Princeton: Princeton University Press.
May, R.M. (in press). Patterns of species abundance and diversity.
May, R.M., Conway, G.R., Hassell, M.P. and Southwood, T.R.E. (in press). Time delays, density dependence and single-species oscillations. *J. Anim. Ecol.*
Orians, G.H. (1973). A diversity of textbooks: ecology comes of age. *Science*, 181, 1238-9.
Oster, G. and Takahashi, Y. (in press). Models for age specific interactions in a periodic environment. *Ecological Monographs*.
Pennycuick, C.J., Compton, R.M. and Beckingham, L. (1968). A computer model for simulating the growth of a population, or of two interacting populations. *J. theor. Biol.* 18, 316-29.
Preston, F.W. (1948). The commonness, and rarity, of species. *Ecology*, 29, 254-83.
Preston, F.W. (1962). The canonical distribution of common-

ness and rarity. *Ecology,* 43, 185-215, 410-32.
Simpson, E.H. (1949). Measurement of diversity. *Nature, Lond.,* 163, 688.
Terborgh, J.W. (in press).
Usher, M.B. (1972). Developments in the Leslie matrix model. *Mathematical Models in Ecology* (Ed. J.N.R. Jeffers), pp.29-60. Oxford, London, Edinburgh and Melbourne: Blackwell Scientific Publications.
Webb, D.J. (in press). The statistics of relative abundance and diversity.
Whittaker, R.H. (1970). *Communities and Ecosystems.* New York: Macmillan.
Whittaker, R.H. (1972). Evolution and measurement of species diversity. *Taxon,* 21, 213-51.

Part one: Studies within a trophic level

The analysis of discrete time cycles

MARK WILLIAMSON

Department of Biology, University of York

INTRODUCTION

The best known cycles in animal numbers are the ten-year cycles in Canada. Moran (1953) examined the correlogram of the number of lynx (*Felis (Lynx) lynx canadensis*) trapped in the Mackenzie River district, and showed that the population had a statistically significant cycle, averaging a little less than ten years from peak to peak. Other Canadian populations of lynx and other animals also exhibit a ten-year cycle. In Labrador, and elsewhere, there has been a well marked four-year cycle in the numbers of fox (*Vulpes fulva* , or, *V. vulpes*) and other organisms (Elton, 1942). Here too the statistical significance of the cycles in the number of pelts is quite definite (Webber and Williamson, in preparation). All these cycles involve animals that live for more than one year, living in a strongly seasonal climate, and having a fairly definite breeding season. A realistic continuous time model of such a situation would have to be quite complicated, allowing for the seasonal cycles. It should be much easier to get a satisfactory discrete time model, giving, in the first place, a prediction of numbers in each year. Moran (1953) fitted a second order auto-regressive model, which is a discrete time model, though with only moderate success.

The intention of this paper is to increase our understanding of the sort of cycles in model systems that can be produced in discrete time, if one allows for the age structure of the population. On the one hand, this might allow a

Ecological Stability
edited by M.B. Usher and M.H. Williamson.

better understanding of natural cycles such as four-year and ten-year cycles. On the other hand, as using the age-structure of populations is one of the easiest ways to produce models with cycles, it is of interest in itself to understand more about the cycles so produced. The system used here was briefly mentioned in Williamson (1967), and is described more fully below. Although strong and regular cycles are easily produced in this way, the length of the cycles differs in some unexpected ways between variants of the system, and so another objective of the study reported here is to try to understand the mathematical conditions that lead to cycles of a particular length.

MATRIX MODELS

The method of producing cycles mentioned in Williamson (1967) and elaborated in this paper may be described as 'matrix-jump'. An ordinary Leslie matrix system, as is now well known, consists of a vector describing the age structure of the population, and a square transition matrix. The premultiplication of the vector by the transition matrix gives a vector, and so the population structure, at the next point in time. Eventually, the population settles down to exponential change in a stable distribution. The rate of population change is given by the dominant latent root (eigenvalue) of the transition matrix, while the dominant vector (eigenvector) gives the stable age distribution. The system can represent an exponential increase or an exponential decrease in the population, in both cases eventually in the stable age distribution, depending on whether the dominant root is greater or less than one. This is equivalent to having a positive or a negative intrinsic rate of natural increase (Williamson, 1972). If now we use two Leslie matrices, the first with a root greater than one and the second with a root less than one, and we choose which matrix to use at any particular instance of time from some property of the population vector, then it is easy to generate population cycles. Fig.1 shows the system mentioned by Williamson (1967). The matrices involved are

$$\begin{bmatrix} 0 & 9 & 12 \\ 1/3 & 0 & 0 \\ 0 & \frac{1}{2} & 0 \end{bmatrix} \quad \text{and} \quad \begin{bmatrix} 0 & 9 & 12 \\ 1/84 & 0 & 0 \\ 0 & \frac{1}{2} & 0 \end{bmatrix}$$

where 9 and 12 represent the net fertilities of mid-aged and old individuals, and the sub-diagonal terms the survival from young to mid-aged and from mid-aged to old.

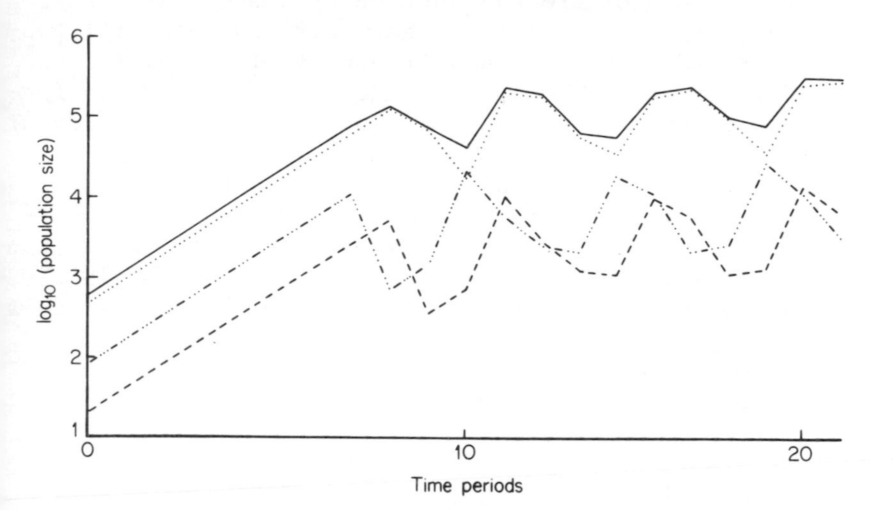

Fig.1. Stable limit cycles with 3 x 3 Leslie matrices, using the system described in Williamson (1967). On the left is a phase of exponential increase. The solid line gives the total population size, the dotted line the number of young, mixed dots and dashes mid-aged and dashes old-aged. The system produces cycles between four and five units long

The only difference between the matrices is the survival from young to mid-aged, but the left-hand matrix has a dominant root of +2, while the right-hand matrix has one of $+\frac{1}{2}$. If the left-hand matrix is used when the population, or part of it, is below a certain size, while the right-hand is used when the population is above that size, oscillations are produced as shown in Fig.1. At the left-hand side the population can be seen to be increasing exponentially, doubling each time interval, and in the stable age distribution of 24 young to 4 mid-aged to 1 old. Once the population reaches the critical value, in this case of 1,000 old, the population goes into cycles, and is never thereafter in the stable age distribution appropriate to either transition matrix. All three age classes and the total population size oscillate.

The fact that such a system can produce an oscillation is not particularly surprising and somewhat more mathematically sophisticated oscillations have been described by Pennycuick *et al.* (1968) and Usher (1972). Marked changes in survival and fertility with density are known in some species, Nicholson's (1957) blowflies being the best known, though these changes are no doubt never exactly step functions. Although it is obvious that these systems will produce cycles, the amplitude and frequency of the cycles so produced is not so obvious. An understanding of these features, and a comparison with the amplitude and frequency of real cycles, might lead to a better understanding of the behaviour of real populations. There are a number of methods available for the study of real data, and by testing these on artificial data, it should be possible to gain some idea of which are likely to be the most useful in practice. So the two topics dealt with in this paper are a discussion of the types of cycle produced by the matrix-jump system, and a discussion of the results produced by analysing such cycles by some standard methods.

MATRIX-JUMP SYSTEMS OF ORDER TWO

Because the results of using matrix-jump systems of order three, that is, using three by three matrices as shown in Fig. 1, gave unexpected variations in the time from one peak population to the next, I thought more progress might be made by studying the smallest possible Leslie matrices, namely those two by two. As before, two matrices were used in any one computer run, one with a dominant root greater than one, and one with a dominant root less than one. For convenience, I refer to these as the 'up' and 'down' matrices. The matrices used are shown in Table 1. Only one 'up' matrix has been used while there are four different

Table 1. The five matrices used in the simulations

Matrix name	U	D_p	D_f	D_m	D_x
Latent root	+2	$+\frac{1}{2}$	$+\frac{1}{2}$	$+\frac{1}{2}$	+1/5
Matrix	$\begin{bmatrix} 0 & 16 \\ 1/4 & 0 \end{bmatrix}$	$\begin{bmatrix} 0 & 16 \\ 1/64 & 0 \end{bmatrix}$	$\begin{bmatrix} 0 & 1 \\ 1/4 & 0 \end{bmatrix}$	$\begin{bmatrix} 0 & 4 \\ 1/16 & 0 \end{bmatrix}$	$\begin{bmatrix} 0 & 4 \\ 1/100 & 0 \end{bmatrix}$

'down' matrices. Three of the 'down' matrices are balanced against the 'up' in a sense that the dominant root of the 'up' is 2, while that of the three 'down' ones is $\frac{1}{2}$. The fourth 'down' matrix produces a more rapid decrease in the population, as it has a dominant root of $1/5$. In all cases the signal used when changing from the 'down' to the 'up' matrix, and vice-versa was the total population size. All these matrices are of the form

$$\begin{bmatrix} 0 & a^2 \\ b^2 & 0 \end{bmatrix}$$

and such a matrix has two roots, $+ab$ and $-ab$. The two roots are of equal modulus, and so although a stable age distribution can be defined, which is given by the vector associated with the root $+ab$, the population will not approach this distribution if it does not start in it, but will oscillate between two age distributions, one of them being that from which it started. This phenomenon, and indeed the whole topic of Leslie matrices, was first described by Bernardelli (1941) and it is convenient to describe waves in the population structure resulting from having a Leslie matrix with roots of equal modulus, as 'Bernardelli waves'. The types of Leslie matrix that have roots of equal modulus are discussed fully by Sykes (1969). Bernardelli waves take place during the exponential growth of the population, and are meta-stable, not returning exactly to their previous form if disturbed, while the waves produced by the matrix-jump system are stable limit cycles.

Multiplication of pairs of the matrices shown in Table 1 lead to matrices of the form

$$\begin{bmatrix} a & 0 \\ 0 & b \end{bmatrix}$$

and such matrices have roots a and b. It is possible for a and b to have the same modulus, and so some of the systems described below involve Bernardelli waves superimposed on stable limit cycles. Fortunately this in no way complicates the interpretation of the cycle. If it had, a small term in the second row in the second column of the matrix, representing some adult survival, and a negative exponential adult life table (Williamson, 1959) could have been used.

Both because it represents what happens in real populations, and because it fits one of the theoretical points below, most of the runs were done with the population represented as integers, rather than as fractional numbers. This was achieved by rounding off the fractional parts after multiplication by the matrices, but it could equally well be regarded as making small alterations to the elements of the matrices at each multiplication, so that the result was an integer one. Considering the operation this second way leads to a simpler interpretation of population cycles, as will be seen, though in practice the difference between the simulation done in integers and one done in real numbers is not detectable on a graph.

THE THEORETICAL LENGTH OF MATRIX-JUMP LIMIT CYCLES

Before considering the detailed results from the simulations, there are three theoretical points that need to be made. A fourth point is deferred until the last section ('The causes of quasi-cycles'). All of these relate to the length of the limit cycle. The first is that, as I have been working in integers, the set of possible population sizes that can be reached during the limit cycle is finite. This set consists of two sub-sets, above and below the signal. Below, in this case below a total population size of 1,000, it is easy to see that the sub-set is finite. Even though with, say, a three by three matrix, the same total population could be made up in a number of different ways, these ways are themselves finite, and so the sub-set of populations that total 1,000 is also finite. With the two by two matrices, there is in fact no difference in the size of the set whether one considers the total population size or the individual age classes. Above the signal, all populations that can be included in the limit cycle have to be reached from a population below the signal. Because of the rules of the system, each population structure below the signal can only give rise to a finite number of populations above the signal, and of course most populations below the signal give rise to none above. So the sub-set above is also finite, and thus the whole set is finite. Each point in the whole set represents a unique age structure. Working with fixed rules, the transitions from any one point will always be to one, and only one particular, other point Consequently, the maximum length of the cycle occurs if eac point in the set occurs once in the cycle; no point can occur twice, because a second occurrence is the start of a new cycle by definition. So the total length of the limit cycle must be less than or equal to the total of the set. This proposition is similar to that which shows that pseudo random numbers on a computer will repeat eventually. However, the cycles implied by this proposition could be of very considerable length, will vary in length depending on the signal used, and there could be cycles of different lengths coexisting within the same system. The system will certainly have crossed the signal value many times during the course of one complete cycle. This introduces one of the main problems in analysing the short cycles seen in the matrix-jump system: the full cyclic system may have a very long period, composed of much shorter and more obvious quasi-cycles. If such a system operated in a real population, it is the quasi-cycles that would be noticed by an ecologist, because the full cycle would, even with the best conceivable data, be too long for him to observe.

The second and third theoretical points suggest somewhat shorter cycles. The operation of the 'up' and 'down' matrices is a continued product, which may be represented as

$$(\mathrm{U.U.U...D...U...}) = \mathrm{J}$$

The system would eventually find a J in which the dominant root λ_1 is equal to one. The roots of a matrix product are unchanged by any cyclic rearrangements (see for instance Barnett and Storey (1970), p.20) so if the cycle is n terms long, there will be nJs, each starting from a different point in the cycle, all with $\lambda_1 = 1$. In finding such a J, quasi-cycles using shorter sequences, in which λ_1 almost equals one, may also be found.

The third theoretical point is a simple elaboration of the second. Instead of writing the system as a continued product, it can be written as a large matrix with all the individual 'up' and 'down' matrices along the principal sub-diagonal, with one in the top right-hand corner. For instance, if the product UUDD is the appropriate one, this can be written as a matrix

$$\begin{bmatrix} 0 & 0 & 0 & \mathrm{D} \\ \mathrm{U} & 0 & 0 & 0 \\ 0 & \mathrm{U} & 0 & 0 \\ 0 & 0 & \mathrm{D} & 0 \end{bmatrix}$$

If the product matrix has a dominant root equal to one, so will the expanded matrix. As before, if there are shorter product matrices with roots almost equal to one, so there will be smaller expanded matrices with roots almost equal to one, and these will be important in the behaviour of the system. In this expanded representation, the dominant vector gives the age structure at each point round the limit cycle. This vector also defines the point in hyperspace, so this representation maps a limit cycle as a point As it is easier to determine the behaviour of a system in a neighbourhood of a point, rather than round a cyclic locus, this method of thinking of limit cycles may be useful.

These three theoretical points all show that there will be a limit cycle, but they give no indication of its length, and suggest that there may well be quasi-cycles within the full cycle. I will leave the consideration of these quasi-cycles until the last section, so that the results of the simulations may be considered first.

RESULTS OF THE SIMULATIONS

With one U matrix and four D matrices there are four simulations to be considered, each one using the U matrix and only one of the D matrices. The matrices have already been given in Table 1. The U matrix has a dominant root of +2. D_p differs from U only in the p term, D_f in the f term, and D_m is a mixed modification of U involving both terms. All three have a dominant root of $+\frac{1}{2}$. The fourth 'down' matrix D_x, with a dominant root of $+1/5$, is a modification again of D_m.

The results of the first three D matrices are so simple as not to need illustration. D_m produces an alternating sequence of U and D and the population curve is a simple saw-tooth shape. D_p also produces a fairly simple system, this time with a four-point cycle rather like the battlements on a castle with the population size alternately high twice and then low twice. The matrix sequence that produces this is $(UUDD)^n$. Only every other U and D produces an appreciable change in the total population size, the other two changing the age structure rather than the total population. The third system, using D_f, also produces a four-point cycle, though a slightly more interesting one. In this case the population size goes from high to mid to mid to low and then back to high again. The sequence in th product is $(UDDD)^n$, and it is interesting to note that ther are three times as many D's as U's. Nevertheless it is eas to show that the dominant root here is still +1. The difference is that in the two preceding cases the other root of the product is −1, while in the third case it is $+^1/_{16}$. So while the first two systems stay in a Bernardelli wave, with two non-interacting population classes, in this third system one of these two classes becomes extinct. After the system has run for a while, there is only one age class in the population.

In none of these cases does the population approach the dominant latent vector of either the $\mathbf{U}$ or the $\mathbf{D}$ matrix, and indeed from the second theoretical point made above, there is no reason to suppose they should. It is well known that there is no necessary relation between the latent roots and vectors of the product $\mathbf{AB}$ to the roots and vectors of $\mathbf{A}$ and $\mathbf{B}$. The consequence of this is that although the population is acted on by two matrices whose 'intrinsic rates of increase' are easily calculated, the population never changes at these rates.

The fourth system using $\mathbf{U}$ and $\mathbf{D}_x$ gives a somewhat more life-like result. A typical run is shown in Fig.2a. Counting from peak to peak, the cycles are three to four

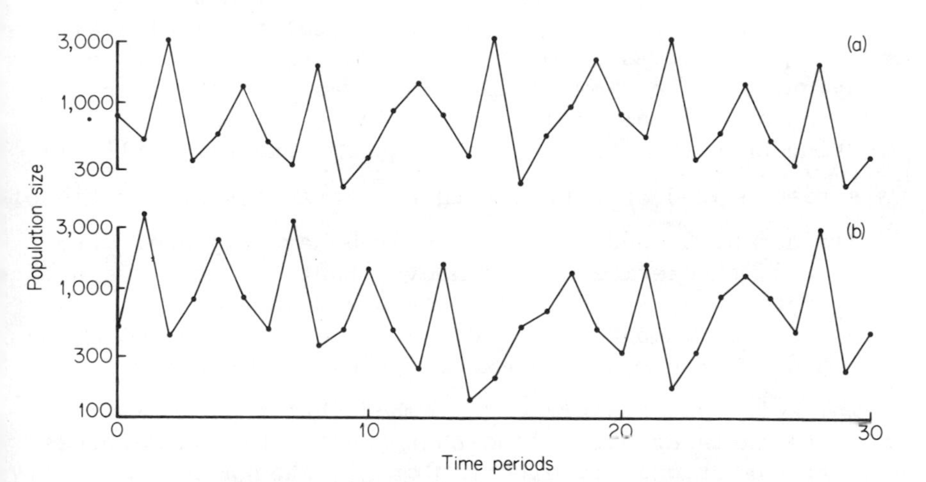

Fig.2. Results of the matrix-jump system using matrix $\mathbf{U}$ and matrix $\mathbf{D}_x$. Top graph (a) the deterministic result. Bottom graph (b) with some stochastic variation

periods long, while from trough to trough they are three or five periods long. There is almost exact repetition after twenty points, though this is still a quasi-cycle as will be shown more clearly below. The sequence in the twenty point quasi-cycle is

D.U.U.D.U.U.D.U.U.D.U.U.U.D.U.U.D.U.U.U.
3 3 3 4 3 4

In this, two D's are usually separated by two U's; the sequence DUUD occurs four times. The pattern of change of the total population size is rather different, and the corresponding sequence is, using *F* for a fall in population size and *R* for a rise

F.F.R.F.R.R.F.F.R.F.R.R.R.F.F.R.F.R.R.R.

In this sequence there is only one instance of two rises between a fall, the sequence (*FRRF*). A simple minded correlation between the DU series and the *FR* series gives a correlation of only +.72, which is certainly formally significant, but perhaps surprisingly low for an entirely deterministic process of this sort. The reason the changes in total population size do not follow the changes in the transition matrices causing them is, again, because the population is never in the stable age distribution appropriate to either of the two transition matrices. This has the important implication, that the total population size and the derivatives from it is insufficient to analyse the behaviour of a complex population, even when it is acting independently of all other populations.

Using only the two matrices U and D_x exactly, will always give a product matrix with a root involving a ratio of 2^n divided by 5^m and this can never be exactly one. From the second theoretical point above, that the matrix sequence will approach a set whose product has a root of one, it might then be thought that this particular system of U and D_x could never produce an exact cycle. This is the point at which it is necessary to remember that the simulations are done in integers, and so that the transition matrices used are never exactly either U or D_x, though they are very close to them, and so the root of the product is not in fact limited in this way. Nevertheless the sequence of 20 U's and D_x's does not have a root of exactly one, its dominant root is 1·0485, which is very close to the shift in total population size produced at the end of the twenty point cycle.

None of the three theoretical points holds exactly if one allows for stochastic variation in the population. Nevertheless the same type of population variation will be seen, and this is shown in Fig.2b. In this simulation, a random variable, distributed rectangularly from 0•75 to 1•25, has been included in each step of the process. It can be seen that the effect is to make the pattern somewhat less regular, both in amplitude and in frequency, and to make the variance in population size larger. As the size of the random variables used in this run is fairly small, the same pattern of population variations can be identified in Figs.2a and 2b. Fig.2b is slightly ahead of Fig.2a at the left of the diagram, and in phase at the right.

As the comparison of the DU series and the *FR* series above has shown it is essential to consider the age structure of the population when analysing the cycle. One advantage of using a two by two matrix is that the oscillations in the age classes can be shown quite simply on a graph, and this has been done in Fig.3. Along the abscissa (x-axis) are the number of young, and along the ordinate (y-axis) are the number of old. So each point on this graph represents a population structure. The signal used in the simulations, total population size equals 1,000 is also shown, and comes out as a curved line as this is a logarithmic plot. To the left and below the signal line, the U matrix is used, and to the right and above D_x. Although the U matrix has a dominant root of +2, and the D of $+1/5$, the population only changes at those rates when it is in the structure given by the dominant vector, the stable age distribution. With other age distributions, either matrix can produce either an increase or a decrease in the total population size. The zones in which these two matrices can act this way are also indicated.

The population sequence in Fig.3 matches that in Fig.2a, in that the start point of Fig.3 is the same as the starting point used in Fig.2a, though Fig.3 only covers the twenty point cycle, while Fig.2a repeats the first half of the cycle. The major type of variation in Fig.3 is a three point cycle, with an apex pointing north-west, separated by a longer cycle involving a zig-zag from south-west to north east. After twenty steps, the population structure has almost, but not quite, returned to its starting point. This system is in fact one with a Bernardelli wave, so that the detailed pattern of the twenty point quasi-cycle depends on the starting point, but the same features are in fact recognisable irrespective of the starting point. It can also be shown that this cycle is reached from any point in the age structure space outside the area of the limit cycle.

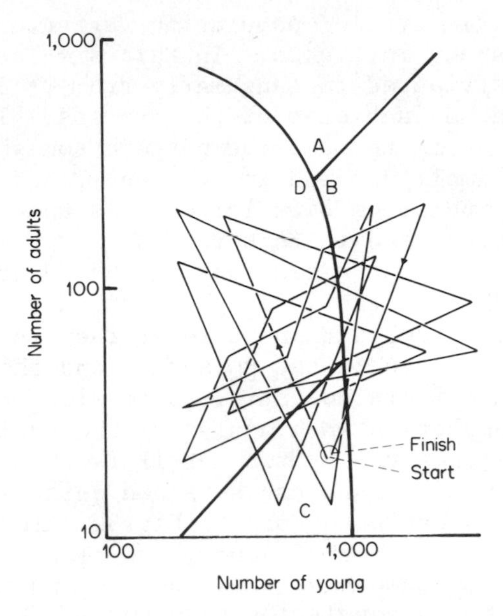

Fig.3. The oscillations in Fig.2a shown as a plot of the number of adults against the number of young. The tendency of the population to go in a triangular path on this diagram, producing a three-year cycle, and the production of an almost exact twenty-year cycle are evident. The line between zones A and B and zones C and D is the signal of total population size = 1000. The matrix $\mathbf{D}_x$ is used in zones A and B; in zone A it produces an increase in the population, in zone B a decrease. Matrix $\mathbf{U}$ is used in zones C and D producing a decrease in population size from zone C and an increase from the zone D

THE ANALYSIS OF THE 20 POINT CYCLE

Although a population cycle of this sort can only be fully analysed knowing the age structure of the population, much of the data that is available for the study of cycles refer only to total population size. It is therefore interesting to try using the methods that might be used on real data on the points of Fig.2a, in order to see how useful these methods are in analysing a system whose origin is known. The methods that might be used in a real system are correlogram analysis, an analysis using a Moran-Ricker diagram (Williamson, 1972) and periodogram analysis. On these data the periodogram analysis does not provide results appreciably different from the correlogram analysis. This is hardly surprising as the autocorrelation, which is plotted in the correlogram, is used in calculating the spectrum used in the periodogram analysis. So this method will not be discussed here.

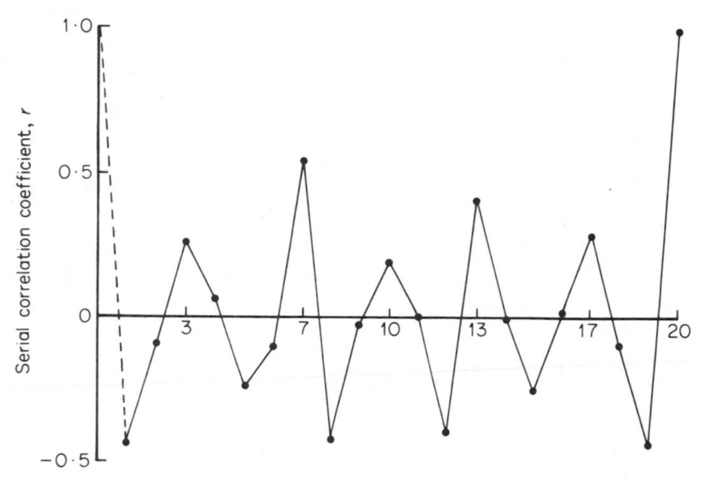

Fig.4. The correlogram of Fig.2a. The quasi repetition at twenty units can be seen. The tendency for a three-year cycle to be the most conspicuous feature of the data is reflected in the prominent peak at a lag of three

The correlogram of the quasi-cycle is shown in Fig.4. The strength of the serial correlation of lag 20 is a dominant feature, but the tendency to three-year cycles is also well marked. The correlogram shows the nature of the cycle clearly enough, but gives no indication of the underlying causes. Correlograms can also be calculated for the runs with stochastic variation such as that shown in Fig.2b. An interesting ratio is that between serial correlations, with the same lag, from deterministic and stochastic runs. Provided the stochastic variation is not too large, this ratio declines progressively with increasing lag. In the run shown in Fig.2b, for instance, the quasi-cycle has taken 21 steps to complete. Consequently with a set of such runs, the serial correlation of lag 20 is much smaller than that shown in Fig.4, the size of the decrease depending on the size of the stochastic variation added. The correlogram, and even more the spectral analysis that can be derived from it, is a satisfactory way of determining the frequency in real data, but is not much help in determining the structure of the series.

Nevertheless the correlogram (Fig.4) is appreciably more helpful than a Moran diagram (Fig.5). As the underlying system involves two matrices, with dominant latent roots

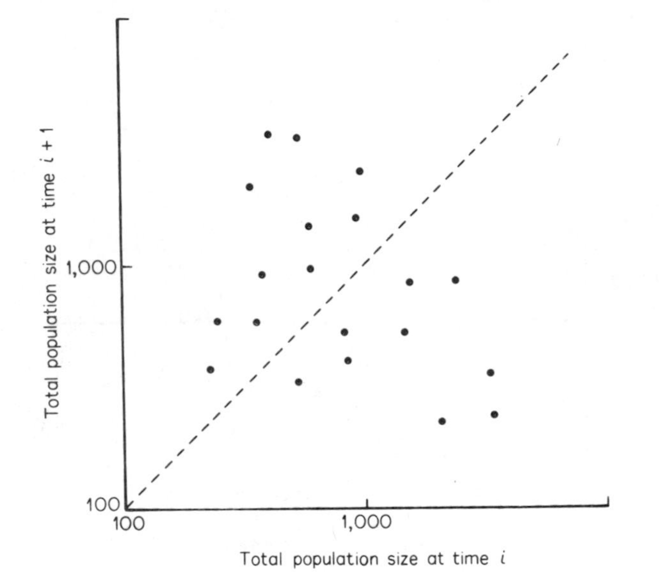

Fig.5. A Moran-Ricker diagram for Fig.2a. While this diagram indicates a tendency to an oscillation, the nature of the cycle, or even its standard length, cannot be deduced

respectively greater than and less than one, and with the switch from one to the other at a definite signal, the system can be visualised in terms of a Moran diagram in an appropriate space. But plotting the results on a Moran diagram in ordinary population-size space does not show this at all, as can be seen from Fig.5. The failure of the points that fall around any sort of line makes it impossible to analyse the cyclic variation in this case. This is also true in practice with the ten-year cycle of the lynx in Canada. My conclusion is that the Moran diagram is an important theoretical aid in understanding how cycles may be produced, but is of relatively little use in practical analysis.

There is in fact only one standard analytical method which could be immediately informative in this case, and that is the k factor analysis of Varley and Gradwell (1960).

However, this requires a knowledge of the age structure of the population. If this knowledge is available, the elements of the matrices are produced at once, and the whole system is clear. If we are to understand the dynamics of complex populations, it is necessary to gather information on their age structure. So far, rather little appropriate data, collected over only a few years, is available for any cyclic species.

THE CAUSES OF QUASI-CYCLES

There is, however, one more aspect of the data in Fig.2 which needs discussion; the fourth point, additional to the three points in the section on 'The theoretical length of matrix-jump limit cycles'. This is the reason for the rather short length of each increasing and decreasing phase in population size. In the system, there is a tendency to a three-year cycle and no individual pair of up and down phases, which would usually be called a cycle, takes longer than five periods. The explanation for this is simple. Under the rules used for the simulation, it is impossible for the population to get very far away from the signal, once it has approached the limit cycle. That is, the operation of a 'down' matrix on any population which is above the signal, will not send it all that far below the signal, and correspondingly the operation of the 'up' matrix from any point below cannot send the population far above the signal. The result is that the population cannot increase or decrease for very long without crossing the signal. In Fig.3, starting from the point at the south-west corner, it is easy to see that three operations of the U matrix have sent the population across the signal. The total population size does not of course necessarily decrease in a matrix-jump system once the signal has crossed, though it does in the example, because the limit cycle does not reach zone A of Fig.3. It is not just the limited ability of either matrix to move the population away from the signal that limits the length. If, for instance, a different D matrix was used, with an even greater rate of decrease, then the population would descend more rapidly, and would take longer to increase back across the signal. But, roughly, the extra time needed to increase would be compensated by the decreased time on the decrease phase. The effect would be to make the cycle more asymmetrical, but not longer. In this matrix-jump system the maximal length from peak to peak, or from trough to trough, depends on the order of the matrices used, rather than on their numerical properties. The amplitude of these cycles can of course be affected by using different matrices of the same order. With two by two matrices, one is limited to something near a four-year cycle, with two declines and two increases of population as a typical pattern. Larger matrices would allow longer cycles. The same conclusion applies even if one uses the more elegant functional variation of the elements of the matrices in relation to the signal, as in Pennycuick *et al.* (1968), or Usher (1972). In these cases, as well as in the matrix-jump systems, the signal from the present population size has been used to produce an immediate effect. With some extra delay in the cycle, and with small matrices, longer population cycles could be produced. So a cycle such as the ten-year cycle requires either complex populations with roughly four biologically distinguishable age classes, or an appreciable delay to the system over and above that which one would naturally expect with a matrix of order 4.

ACKNOWLEDGEMENTS

Nigel Bath, John Byrne and Martin Joyce helped with computation, graphs and comments. Ian Pyle pointed out the relationship with pseudo-random numbers. John Currey made many useful comments on the text, and Joel Cohen on Leslie matrices in general. I am grateful to all of them for thei help. This work was supported by grant B/RG/33423 from the Science Research Council.

SUMMARY

An analysis of the cycles in population size obtained by using a mixture of two Leslie matrices is presented. The cycles produced are complicated, and an exact repetition only takes place after a very considerable time. The major features, quasi-cycles, are most simply explained in terms of the size (order) of the Leslie matrices. For analysing real populations, correlograms are of some use, Moran diagrams of relatively little use, but a full analysis require the knowledge of age structure of the population.

REFERENCES

Barnett, S. and Storey, C. (1970). *Matrix methods in stability theory*. London: Nelson.

Bernardelli, H. (1941). Population waves. *J. Burma Res. Soc.*, 31, 1-18.

Elton, C. (1942). *Voles, mice and lemmings. Problems in Population Dynamics*. Oxford: Clarendon Press.

Moran, P.A.P. (1953). The statistical analysis of the Canadian lynx cycle. I. Structure and prediction. *Aust. J. Zool.*, 1, 163-73.

Nicholson, A.J. (1957). The self-adjustment of populations to change. *Cold Spring Harbor Symp. Quant. Biol.*, 22, 153-73.

Pennycuick, C.J., Compton, R.M. and Beckingham, L. (1968). A computer model for simulating the growth of a population, or of two interacting populations. *J. theor. Biol.*, 18, 316-29.

Sykes, Z.M. (1969). On discrete stable population theory. *Biometrics*, 25, 285-93.

Usher, M.B. (1972). Developments in the Leslie matrix model. *Symp. Brit. Ecol. Soc.*, 12, 29-60.

Varley, G.C. and Gradwell, G.R. (1960). Key factors in population studies. *J. Anim. Ecol.*, 29, 399-401.

Webber, M. and Williamson, M. (in preparation). The four-year cycle in Labrador.

Williamson, M. (1959). Some extension of the use of matrices in population theory. *Bull. math. Biophys.*, 21, 261-3.

Williamson, M. (1967). Introducing students to the concept of population dynamics. *Symp. Brit. Ecol. Soc.*, 7, 169-76.

Williamson, M. (1972). *The analysis of biological populations*. London: Edward Arnold.

Red tides and algal strategies

T. WYATT

Fisheries Laboratory, Lowestoft

Several models have emphasized the importance of turbulence in relation to the dynamics of plankton distribution. Two very simple but instructive examples are those of Kierstead and Slobodkin (1953) in which the size of a plankton patch is determined by the magnitude of the turbulent diffusion, and of Margalef (1967) in which the flushing rate of an estuary plays a similar role. In both these models the potential increase in plankton populations through their reproductive activities is offset by turbulence or advection, and the latter processes effectively determine the minimum reproductive rates of the populations if they are to remain in existence. Margalef's model has been used to interpret data from an Indian estuary in which there is a marked seasonal change in flushing rate as a result of the southwest monsoon (Wyatt and Qasim, 1973). This study suggested that increased reproductive effort is not the only possible response to the problem posed by the model, and that other strategies can be adopted to circumvent the losses consequent on turbulent processes. In the model presented here, loss of reproductive effort through turbulence is again central, though not made explicit in the mathematical formulation.

Dinoflagellates are most frequently responsible for outbreaks of red tide. The blue-green alga *Trichodesmium* is the only other organism commonly involved, though diatoms and ciliates occasionally produce red-tide conditions. A number of features distinguish dinoflagellates from other important groups of phytoplankton. They are often large

Ecological Stability
edited by M.B. Usher and M.H. Williamson.

compared with other flagellated algae, so that their swimming speeds are greater. *Ceratium, Prorocentrum* and *Gonyaulax* can swim 5 to 10m in 12 hours (Peters 1929, Hasle 1950). Like other flagellates, they often possess eyespots and it may be that the design of their flagellae is such that they can orientate themselves to light more effectively than other algae. Many dinoflagellates produce substances which when released into the water are toxic to animals. There is a general feeling, though little data is available, that dinoflagellate reproductive rates are relatively low. One would certainly expect this to be true of the larger species. Finally, as their names sometimes indicate, dinoflagellates are often luminescent, though the value of this feature is not known.

Trichodesmium is also motile in the sense that, like other planktonic blue-green algae, it can produce gas vacuoles which enable it to adjust its level in the water column. Other blue-green algae are responsible for conditions analogous to red-tide in freshwater lakes, and these also possess gas vacuoles. *Microcystis* may be regarded as a freshwater analogue of *Trichodesmium* (cf. Reynolds, 1973) On those rare occasions when diatoms have been found causing red tide, it has been noted that their fat content is very much higher than usual (Tåning, 1951), so that again a mechanism is available for rising towards the surface. The most frequently reported hydrographic and meteorological conditions which precede red tides are heavy rainfall, calm sunny weather, an influx of estuarine waters, and the meeting of dissimilar water masses. Any of these, alone or in combination, might be expected to increase the vertical stability of the water column, either by an increase in the vertical density gradient or by a decrease in the wind-induced coefficient of vertical eddy diffusion. In either case, there will be a reduction in the rate at which phytoplankton is mixed through the upper part of the water column, and dinoflagellates, by virtue of the features mentioned above, may be especially adapted to take advantage o it. The sequence of events which may be generated in such circumstances has been described previously (Wyatt and Horwood, 1973) as follows:
If grazing is constant, then the changes in algal numbers can be expressed by:

$$A_{pt} = A_{p0} \exp\ (R_p - G_{pt})t \qquad (1)$$

$$A_{qt} = A_{q0} \exp\ (R_q - G_{qt})t \qquad (2)$$

$$G_t = G_{pt} + G_{qt} = \text{a constant, C} \qquad (3)$$

where A_{p0} and A_{q0} are the initial numbers of non-motile and motile algae, A_{pt} and A_{qt} are the numbers at time t, R_p and R_q are the respective reproductive rates, and G_{pt} and G_{qt} are the grazing rates at time t. As the radiation declines exponentially with depth, Steele's (1965) equation

$$r = r_m\ (I_z/I_m)\ \exp\ (1 - I_z/I_m)$$

(where r is the reproductive rate at any given depth z, r_m is the maximum reproductive rate, I_z is the radiation reaching depth z, and I_m is the radiation level at which maximum reproduction occurs) may be integrated over a mixing depth d to give a mean reproductive rate R, for the water column of the form

$$R = (2{\cdot}7818\ r_m/dk)\ \left[\exp\{-(I_0/I_m)\ \exp\ (-kd)\} - \exp\ (I_0/I_m)\right] \qquad (4)$$

where I_0 is the radiation level at the surface, k is the extinction coefficient and $I_z = I_0 \exp\ (-kz)$.

When the stability of the water column is such that the motile algae are able to congregate at a particular depth, it may be assumed that they will choose a depth at which the reproductive rate is maximal. This will be when

$$I_m = I_0 \exp\ (-kz)$$

and will generally be near the surface. Hence

$$z = 1/k \ln\ (I_0/I_m) \quad \text{when } I_0 \geq I_m$$

$$z = 0 \quad \text{when } I_0 < I_m$$

Consequently

$$R = r_m \quad \text{when } I_0 \geq I_m$$

$$R = r_m (I_0/I_m)\ \exp\ (1 - I_0/I_m) \quad \text{when } I_0 < I_m$$

Under constant grazing pressure an increase in the abundance of an algal species will result in a fall in the grazing coefficient of that species. If we assume a time lag δt in this effect, then

$$G_{q(t + \delta t)} = \gamma(A_q)^{-1} \tag{5}$$

and from Equation 3

$$G_{p(t + \delta t)} = C - \gamma(A_q)^{-1} \tag{6}$$

Substituting Equations 5 and 6 into Equations 1 and 2, and letting δt be one time step, an explicit relationship is obtained:

$$A_p(t + \delta t) = A_{pt} \exp \left[\{R_p - C + \gamma(A_{qt})^{-1}\}\delta t\right] \tag{7}$$

$$A_q(t + \delta t) = A_{qt} \exp \left[\{R_q - \gamma(A_{qt})^{-1}\}\delta t\right] \tag{8}$$

Equations 7 and 8, given realistic conditions derived from a typical red-tide area, predict an exponential increase in the number of motile algae, and a general fall in the number of non-motile forms (see Wyatt and Horwood 1973). The slope in both cases depends on γ, the density-dependent grazing coefficient, and increases for both A_p and A_q with decreasing values of γ. The model thus suggests a general mechanism for the generation of red tides, and in particular predicts that the causative agents will become more abundant with increasing vertical stability of the water column.

Dinoflagellates are characteristically more abundant during the later stages of marine phytoplankton succession. This is true in particular of those species which give rise to red tides. One possible strategy by which a population can maintain itself in a turbulent environment is by an increase in reproductive effort, i.e. by having a high production (P): biomass (B) ratio (Margalef 1968). The small celled diatoms which initiate succession in temperate seas, and in regions of upwelling, can be assumed to have a high P:B ratio, since they thrive when vertical mixing rate are relatively high. Dinoflagellates found later during succession would then have lower P:B ratios. A similar distinction has been made between r- and K-strategists (MacArthur 1962). In the former, weak density-dependent mortality favours the allocation of relatively more resources to reproduction. Conversely, a stable environment, with more severe density-dependent regulation, will favour increased survival of the individual at the expense of reproductive activity. These are the *K*-strategists. Hairston et al. (1970), who discuss this concept, suggest that "*K* selection" will produce "an increase in body size", the manufacture of a poison, (and) the development of body armor ..." which are characteristic features of dinoflagellates. If dinoflagellates are regarded as *K*-strategists relative to other planktonic algae, then the occurrence of red tides may be seen as the failure of this strategy in the face of conditions more extreme than those for which it was designed. Extreme stability of the water column is a rare event in the marine environment, so that when it occurs those populations most likely to be able to exploit it become unstable, and ultimately destructive.

REFERENCES

Hairston, N.G., Tinkle, D.W. and Wilbur, H.M. (1970). Natural selection and the parameters of population growth. *J. Wildl. Mgmt.*, 34, 681-90.

Hasle, G.R. (1950). Phototactic vertical migration in marine dinoflagellates. *Oikos*, 2, 162-75.

Kierstead, H. and Slobodkin, L.B. (1953). The size of water masses containing plankton blooms. *J. mar. Res.*, 12, 141-7.

MacArthur, R.H. (1962). Some generalized theories of natural selection. *Proc. natn. Acad. Sci. U.S.A.* 48, 1893-7.

Margalef, R. (1967). Laboratory analogues of estuarine plankton systems. *Estuaries* (Ed. by G.H. Lauff), pp.515-21. Publ. Amer. Ass. Advan. Sci. 83.

Margalef, R. (1968). *Perspectives in ecological theory*. Chicago and London: University of Chicago Press.

Peters, N. (1929). Über Orts-und Geisselberwegund bei marinen Dinoflagellaten. *Arch. Protistenk.*, 67, 291-321.

Reynolds, C.S. (1973). Growth and buoyancy of *Microcystis aeruginosa* in a shallow eutrophic lake. *Proc. R. Soc. (B.)*, 184, 29-50.

Steele, J.H. (1965). Notes on some theoretical problems in production ecology. *Mem. Ist. Ital. Idrobiol.*, 18, Suppl., 383-98.

Tåning, A.V. (1951). Olieforurening af havet og massedod af fugle. *Nat. Verd., Kbh.*, 35, 38-43.

Wyatt, T. and Horwood, J. (1973). A model which generates red tides. *Nature, Lond.*, 244, 238-40.

Wyatt, T. and Qasim, S.Z. (1973). Application of a model to an estuarine ecosystem. *Limnol. Oceanogr.*, 18, 301-6.

Part two: Temporal studies between trophic levels: host–parasite systems

Mathematical models of host-helminth parasite interactions

ROY M. ANDERSON

Department of Biomathematics, Oxford University

INTRODUCTION

In ecological texts a wide variety of meanings have been attached to the word stability, the precise implication often depending on the scientific background of the author. Biologists may argue that a stable population is one that has persisted in a given habitat for a long period of time and has neither become extinct nor reached plague proportions. In this framework a precise definition of the time span in which the population has persisted is obviously important. This concept of stability focuses on the ability of a species to survive environmental change and does not refer to fluctuations in time of the numbers of individuals within the population. Thus between the two limits, extinction and plague, the population may oscillate widely but still be regarded as stable due to its continued existence.

A more rigorous approach to the question of stability is possible if a mathematical description of the biological processes under study can be formulated, enabling the numerical stability of the system to be examined. This type of approach is usually adopted by the mathematician or physical scientist who is invariably dealing with a quantitative model describing rates of change of the variables involved. In this type of framework stability usually implies that, if the system is perturbed by small or large displacements, it returns either to the equilibrium state from which it was displaced or to a new equilibrium state. When deal-

Ecological Stability
edited by M.B. Usher and M.H. Williamson.

ing with biological models the perturbations are usually caused by environmental changes. If the system departs from a steady state and fails either to return or find a new equilibrium state then it is regarded as unstable.

Ideally for any biological system it is theoretically possible to construct a description that is a list of the parameters necessary to specify the system completely. These parameters include such things as population rates of change, environmental variables and time. Given that there are n such parameters for a specific biological system, then in the n dimensional parameter space there will be a region A, within which the parameters have values which can occur in the real world. Within this region A there will be another region B (or regions in a non linear system) of parameter values which give rise to equilibrium states for the populations in the system. The space B will contain certain areas which give rise to stable equilibria and other areas producing unstable equilibria. Similarly, if there are p populations in the system there will be regions in the p dimensional population space which are stable and regions which are unstable. The structural stability of the system depends on the region or regions within the parameter space which lead to stable equilibrium in the population space. This concept of stability provides a more precise definition but can only be utilised if a mathematical model of the biological processes under study is successfully formulated.

When investigating host-helminth parasite interactions conclusions about the stability of the system are often difficult to obtain due to the complex nature of parasite life cycles. These cycles may involve a final host and one or more intermediate host populations as well as a number of distant parasite populations formed by different developmental stages in the life cycle which occupy different micro- or macro-environments.

For the parasite life cycle to be completed the appropriate host populations must be present since the majority of helminth parasites are host specific. If one host population becomes extinct the parasites life cycle is terminated and so a set of boundaries are imposed on the stability of the system. Lewontin (1969) termed this type of situation a bounded system, the boundaries being the extinction of any population in the host parasite life cycle. As long as the populations are fluctuating within the boundaries between equilibrium states the system can be regarded as stable. In reality, due to chance or stochastic variation, the populations may never actually be in a steady state, simply fluctuating around the equilibrium.

This concept requires a further modification when applied to parasite life cycles since temporary extinction of say the adult parasites in the final host may not necessarily lead to the extinction of the species in the habitat. This is due to the survival of either the larval parasites in the intermediate host or infective eggs or larvae in a free living state. In these instances a reservoir of infection is provided which can lead to the re-establishment of the adult parasites in the final host. This type of extinction may occur seasonally in a periodic cyclic manner due, for example, to a seasonal change in the feeding or behavioural habits of the host. The boundaries of the stable system in this case need not include the extinction of the adult parasite population, the survival of the host population being the critical factor.

Two approaches towards the mathematical description of helminth parasite life cycles are formulated. The first is stochastic and deals with a distinct compartment of the life cycle while the second is deterministic and describes the whole cycle. Since a mathematical description is attempted, a more rigid physical approach to stability is adopted.

HELMINTH PARASITE LIFE CYCLES

Parasite life cycles are divisable into two groups, direct life cycles involving a single host and indirect cycles involving two or more hosts. The direct life cycle, represented in Fig. 1a, involves two distinct parasite populations occupying separate habitats, the adult parasites in the host and the parasite eggs free living in the aquatic or terrestrial environment. Both populations are basically controlled by immigration and death processes, although an emigration process affects the egg population. No true birth process occurs since the birth of an egg does not directly increase the adult parasite population but becomes an immigrant into the free living parasite egg population. The population processes affecting the host are not described.

The indirect life cycle represented in Fig. 1b is more complicated and involves two host populations and three distinct parasite populations. If transmission is achieved by the ingestion of the intermediate host by the final host then two types of population interaction occur. A host-parasite relationship exists between the final host and

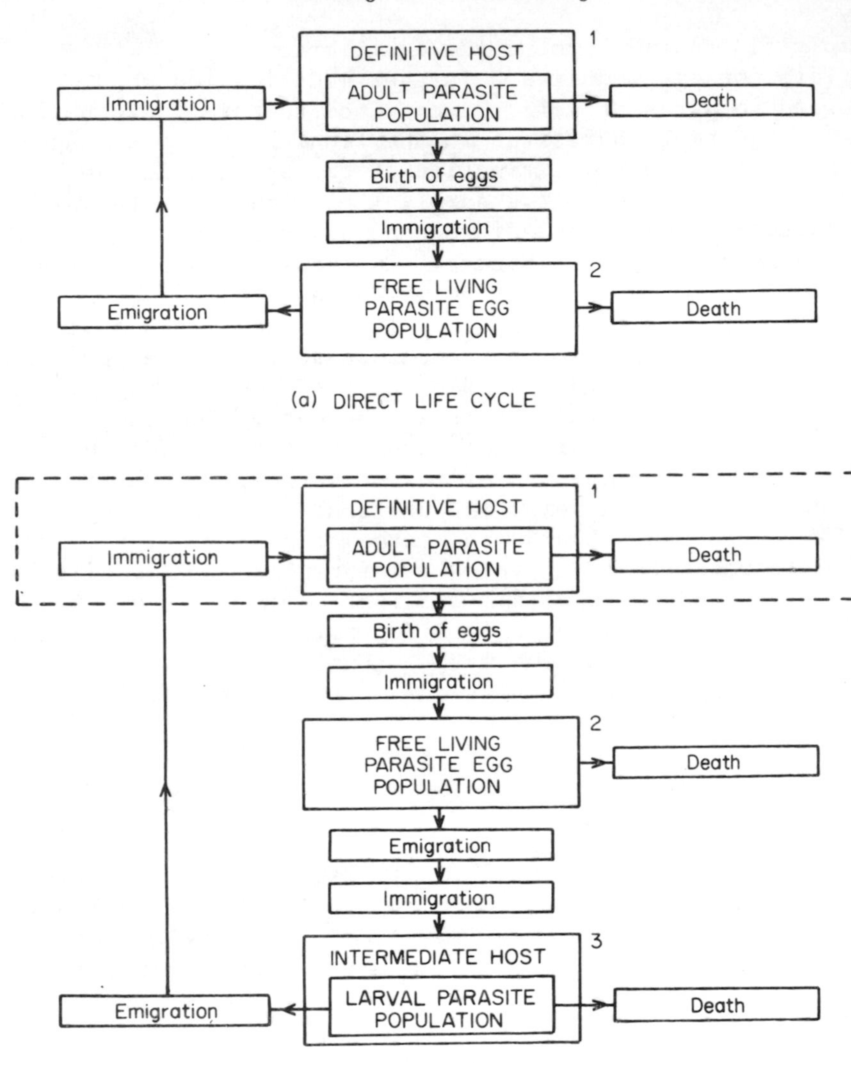

Fig.1. Diagrammatic flow charts of a direct life cycle (a) and an indirect life cycle (b)

adult parasites and also between the intermediate host and larval parasites, while a predator-prey relationship occurs between the definitive and intermediate hosts and, in essence, between the intermediate host and parasite egg population. The system is complicated further by the occurrence of developmental time lags. It is thus possible for the adult or larval parasites to become extinct without the termination of the life cycle occurring. This type of process is exhibited by various fish parasites where the feeding behaviour of the host alters seasonally, resulting in a periodicity of immigration of the larval parasite into the final host (Anderson, 1971; Kennedy, 1970).

When investigating the stability of such complex biological systems, it seems realistic to break the life cycle into compartments (Fig. 1b) and examine these separately before inspecting the complete process. This approach is best illustrated by considering a specific example.

STOCHASTIC MODEL OF AN ADULT PARASITE POPULATION

Caryophyllaeus laticeps (Pallas, 1781), a pseudo-phyllidean cestode of cyprinid fishes in Europe, has an indirect life cycle involving one intermediate host, a tubificid oligochaete. The adult parasite occurs in the gut of the final fish host attached to the intestinal wall by a large scolex and has a life span of up to two months (Kulakowskaja, 1964). Eggs are laid by the mature parasite in the fish gut and pass to the external aquatic habitat with the faeces of the host. These eggs, when ingested by various species of tubificid oligochaetes undergo further development and become infective to the fish after a period of four months (Kulakowskaja, 1962) and can remain infective in the intermediate host for periods greater than one year. The cycle is completed by the ingestion of the infected oligochaete by the fish host, the larval parasites developing to maturity in the fish gut.

The immigration death process of the adult parasite population, compartment 1 in Fig. 1b, is a convenient point to start in the examination of the life cycle. A simple deterministic model to describe the rate of change of the adult parasite population in a single host can be formulated as follows:

$$\frac{dN_t}{dt} = \lambda(t) - \mu(t)\, N_t \qquad (1)$$

where N_t is the number of parasites at time t and $\lambda(t)$ and $\mu(t)$ are respectively the immigration and death rates. These rates of change are known to be seasonal cyclic functions of time (Anderson, 1971).

It is assumed in Equation 1 that the immigration rate $\lambda(t)$ is independent of N_t. This is not strictly true since the availability of infected tubificids will depend on N at some previous time interval. However, since the model attempts to describe simply one compartment of the life cycle this is ignored. The death rate is taken to be density independent, an assumption borne out by the examination of natural populations of *C. laticeps* (Anderson, 1971). No evidence is available concerning the influence of previous experiences of infection and thus this factor is unaccounted for in the model.

A simple model to describe the seasonal cyclic immigration rate $\lambda(t)$ caused by the seasonal feeding behaviour of the host can be formulated as follows

$$\lambda(t) = a + b\left[\sin\frac{2\pi(t-\tau)}{12} + 1\right]$$

where a and b are constants and τ is a phase angle. The immigration rate is known to be a random variable $\Lambda(t)$ due to natural variability between different hosts in the fish population, thus

$$\Lambda(t) = A + B\left[\sin\frac{2\pi(t-\tau)}{12} + 1\right] \tag{2}$$

where A and B are random variables with means m_A, m_B and variances σ_A^2, σ_B^2. Then

$$E\ \{\Lambda(t)\} = m_A + m_B\left[\sin\frac{2\pi(t-\tau)}{12} + 1\right] \tag{3}$$

and if A and B are assumed to be independent

$$\text{Var}\ \{\Lambda(t)\} = \sigma_A^2 + \sigma_B^2\left[\sin\frac{2\pi(t-\tau)}{12} + 1\right] \tag{4}$$

The parameters m_A, m_B, σ_A^2 and σ_B^2 can be estimated by fitting the model to observed data by the technique of least squares (Fig.2a). The death rate of the adult parasite is correlated with water temperature (Anderson, 1971) which varies seasonally. A simple exponential relationship exists between death rate $\mu(T)$ and water temperature T

$$\mu(T) = a\ \exp\ (bT) + c \tag{5}$$

where a, b and c are constants. This non-linear model was fitted to observed data by a least squares technique, using Marquardt's algorithm (Conway, Glass and Wilcox, 1970) (Fig.2b). Since water temperature varies seasonally Equation 5 can be rewritten

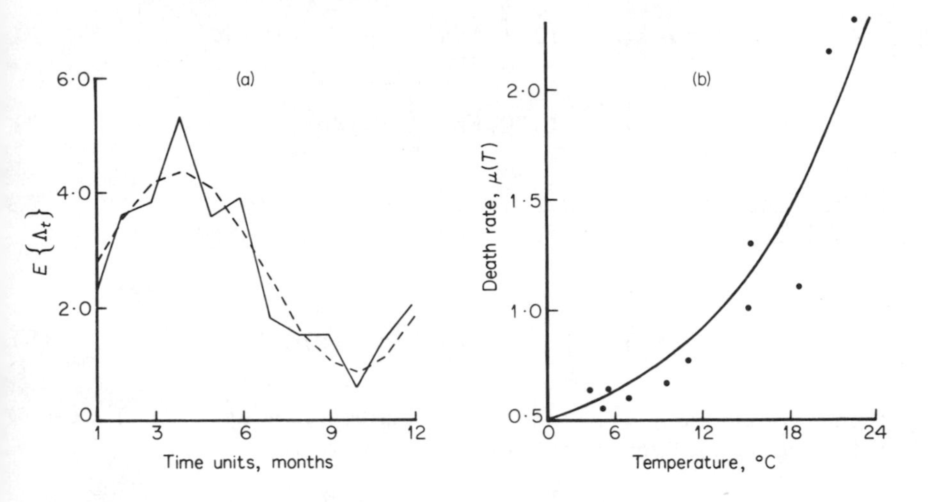

Fig.2. (a) A comparison of observed and expected numbers of immigrant *C. laticeps* gaining entry to a single host in a monthly unit of time (1=February 1969, 12=January 1970). Continuous line represents observed data, dashed line expected. (b) Relationship between water temperature and the death rate of the adult *C. laticeps*

$$\mu(t) = a \exp \left\{b(x + y \sin \frac{2\pi t}{12} + z \cos \frac{2\pi t}{12})\right\} + c \qquad (6)$$

where μ is now a function of time and x, y and z are constants. The modelling of the immigration death process of the parasite population for a single host can be achieved by treating the problem as a Markov process in continuous time, thus entailing a continuous flow of probability distributions. The stochastic treatment of such a non-homogeneous process is described in standard texts on stochastic processes (Bartlett, 1955; Bailey, 1964).

Where $P_n(t)$ is the probability that $N(t)$ takes the value n and the probability generating function $\Pi(z,t)$ is defined by

$$\Pi\ (z,t) = \sum_{n=0}^{\infty} P_n(t)z^n$$

and given the initial conditions $\Pi(z,0) = Z^a$ where a is the parasite population size at time $t = 0$ then

$$\Pi(z,t) = \exp\left[(z-1)\int_0^t \lambda(u)\exp\left(-\int_u^t \mu(v)dv\right)du\right]$$

$$\cdot\left[1+(z-1)\exp\left\{-\int_0^t \mu(v)dv\right\}\right]^a \tag{7}$$

In the case of $a = 0$ or t large then Equation 7 is the probability generating function of a Poisson process with mean

$$E\{N_t\} = \int_0^t \lambda(u)\exp\left\{-\int_u^t \mu(v)dv\right\}du \tag{8}$$

If the adult parasites have a maximum life span of s time units then Equation 8 becomes

$$E\{N_t\} = \int_{t-s}^t \lambda(u)\exp\left\{-\int_u^t \mu(v)dv\right\}du \tag{9}$$

Equation 9 represents the mean number of parasites at time t in a single host. The distribution to which this mean belongs is unobservable since the death of a host is inevitable if counts are to be made of the parasites and each host in the population is different due to varying experiences of infection. However the model can be extended to describe a sample of hosts drawn from a population by treating the immigration rate as a random variable. The death rate is assumed to be deterministic having the same value at one point in time for each parasite.

Incorporating the models for $E\{\Lambda(t)\}$ and $\mathrm{Var}\{\Lambda(t)\}$ represented in Equations 3 and 4, Equation 9 becomes

$$E\{N_t\} = \int_{t-s}^t E\{\Lambda(u)\}\exp\{-\int_u^t \mu(v)dv\}du$$

letting

$$s(u) = \sin\frac{2\pi(t-T)}{12} + 1$$

and

$$g(u) = \exp\{-\int_u^t \mu(v)dv\}$$

then

$$E\{N_t\} = m_A\int_{t-s}^t g(u)du + m_B\int_{t-s}^t s(u)g(u)du \tag{10}$$

where $E\{N_t\}$ is now the mean number of parasites per host at time t. With the assumption that A and B are independent then

$$\text{Var}\ \{N_t\} = \sigma_A^2 \left[\int_{t-s}^{t} g(u)\text{d}u \right]^2 + \sigma_B^2 \left[\int_{t-s}^{t} s(u)g(u)\text{d}u \right]^2 \quad (11)$$

If $P_n(t)$ is now the probability of observing n parasites at time t in any host in the population, working with the generating function $Q(z,t)$ and letting

$$h(t) = \int_{t-s}^{t} g(u)\ \text{d}u \text{ and } k(t) = \int_{t-s}^{t} s(u)\ g(u)\text{d}u$$

then

$$Q(z,t) = \text{E}\{\exp[\{h(t)A + k(t)B\}\ (z-1)]\}$$

Thus where M_A and M_B are the respective moment generating functions of the independent random variables A and B then

$$Q(z,t) = M_A\ \{h(t)\ (z-1)\}M_B\{k(t)\ (z-1)\}$$

If the random variables A and B are assumed to have independent distributions of the gamma form with parameters α, β and $\hat{\alpha}$, $\hat{\beta}$ and moment generating functions

$$M_A(\theta) = \frac{1}{(1-\alpha\theta)^{\beta}} \text{ and } M_B(\theta) = \frac{1}{(1-\hat{\alpha}\theta)^{\hat{\beta}}}$$

then

$$Q(z,t) = [1 - \alpha h(t)\ (z-1)]^{-\beta}\ [1 - \hat{\alpha} k(t)\ (z-1)]^{-\hat{\beta}} \quad (12)$$

It is interesting to note that the generating function $Q(z,t)$ is formed from the product of the generating functions of two negative binomial distributions with parameters α, β and $\hat{\alpha}$, $\hat{\beta}$ and thus the resulting distribution is over dispersed. In the derivation of the generating function one point requires further clarification namely the assumption relating to the distribution of the random variables A and B. Although no biological information is available concerning these distributions, since the gamma is a flexible non-negative continuous distribution which can be fitted to a wide variety of observed distributions, this assumption is not unrealistic and is a natural extension of the gamma form assumption leading to the negative binomial distribution in the case of a single random variable.

The mean and variance of the probability distribution

$P_n(t)$ are derived from the generating function $Q(z,t)$ in the usual manner

$$E\{N_t\} = h(t)\ \beta\alpha + k(t)\ \hat{\beta}\hat{\alpha} \tag{13}$$

$$\mathrm{Var}\{N_t\} = h(t)^2\ \beta\alpha^2 + k(t)^2\ \hat{\beta}\hat{\alpha}^2 \tag{14}$$

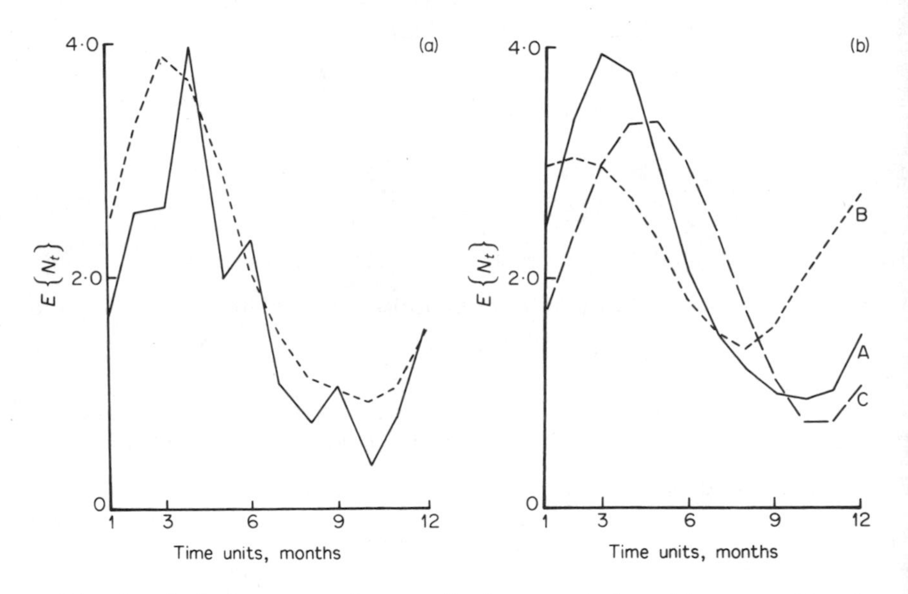

Fig.3. (a) A comparison of observed (continuous line) and expected (dashed line) numbers of adult *C. laticeps* per host. (1=February 1969, 12=January 1970). (b) A comparison of observed and expected numbers of adult *C. laticeps* per host for three different models (A) $dN_t/dt = \lambda(t) - \mu(t)N_t$, (B) $dN_t/dt = \lambda - \mu(t)N_t$, (C) $dN_t/dt = \lambda(t) - \mu N_t$

A comparison of observed and expected values of the mean number of parasites per host for a twelve monthly period is shown in Fig.3a. Comparing Equations 10 and 13 and also 11 and 14, the parameters of the gamma distributions can be derived. By the expansion of Equation 12 it is possible to calculate the individual terms of the probability distribution $P_n(t)$. It is interesting to compare these probabilities with those of a negative binomial distribution with the same mean and variance (Table 1). The theoretical distribution of the immigration death model is slightly different from the negative binomial with larger initial probabilities (i.e. $P(0)$ and $P(1)$ terms) and comparatively larger terms in the tail of the distribution. The tail is thus more drawn out and the distribution more highly skewed although the mean:variance ratio is the same. For practical purposes there is little difference between the two distributions and a significant fit of observed data to the negative binomial distribution is invariably linked with a significant fit to the theoretical distribution derived from the immigration death process. It is interesting to note that the negative binomial has proved to be a good empirical model for observed counts of *C. laticeps* per host (Kennedy, 1969; Anderson, 1971).

Table 1. The comparison of the negative binomial with the theoretical distribution generated by the immigration-death process. In this table E $\{N_t\}$ = 2·07 and Var $\{N_t\}$ = 9·63

No. of parasites	Number of hosts (sample size = 30)	
	Negative binomial	Theoretical models
0	12·73	13·98
1	5·55	5·09
2	3·39	3·01
3	2·27	2·01
4	1·59	1·42
5	1·14	1·04
6	0·83	0·77
7	0·61	0·58
8	0·45	0·45
9	0·34	0·34
10	0·26	0·27
11	0·19	0·21
12	0·15	0·16
13	0·11	0·13
14	0·08	0·10
15	0·06	0·08
16	0·05	0·06
17	0·04	0·05
18	0·03	0·04
19 and over	0·10	0·13

The relative effects of the immigration and death parameters on the stability of the seasonal cyclic behaviour of N_t are examined by considering various types of models.

For example consider the following deterministic models to describe the rate of change in N_t

$$(1) \quad \frac{dN_t}{dt} = \lambda(t) - \mu(t)\, N_t$$

$$(2) \quad \frac{dN_t}{dt} = \lambda - \mu(t)\, N_t$$

$$(3) \quad \frac{dN_t}{dt} = \lambda(t) - \mu N_t$$

Since both $\lambda(t)$ and $\mu(t)$ are cyclic functions of time, each of the above models will generate cyclic behaviour in N_t.

The means of the equivalent stochastic models are as follows

$$(1) \quad E\{N_t\} = \int_{t-s}^{t} \lambda(u) \exp\left\{-\int_{u}^{t} \mu(v)dv\right\}du$$

$$(2) \quad E\{N_t\} = \lambda \exp\left\{-\int_{t-s}^{t} \mu(u)du\right\} \int_{t-s}^{t} \exp\left\{\int_{t-s}^{u} \mu(v)dv\right\}du$$

$$(3) \quad E\{N_t\} = \exp(-\mu t) \int_{t-s}^{t} \exp(\mu u)\lambda(u)du$$

Numerical values for $E\{N_t\}$ for each of the above equations are shown in Fig.3b. The graph illustrates clearly that when λ is constant, model (2), then the temperature dependent death rate has an appreciable effect on the population size of *C. laticeps* causing severe losses during the warm summer months. On the other hand, when μ is constant the mean number of parasites per host sample reflects the seasonal feeding behaviour of the host. The combined effects of $\lambda(t)$ and $\mu(t)$ cause a peak in population size in the late spring and a depression in autumn.

The stability of the model in relation to natural variation in the water temperature of the habitat is examined by simulation studies letting x, y and z in Equation 6 be random variables. Using means and variances of the mean monthly temperatures obtained from a number of years data and assuming these means to be normally distributed, a pseudo-random number generator was used to select a series of values for the mean monthly temperatures and thus to

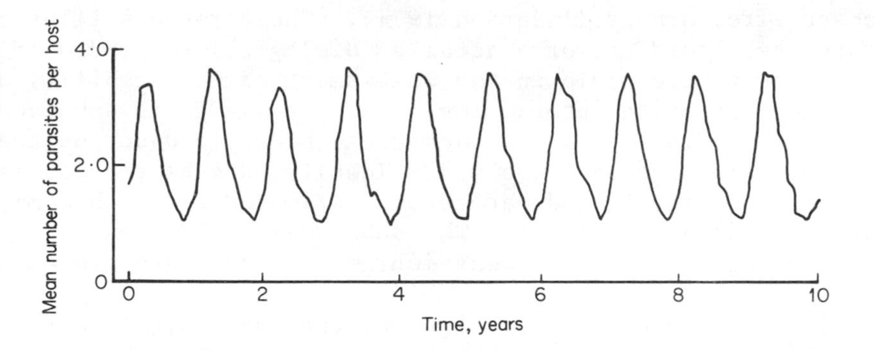

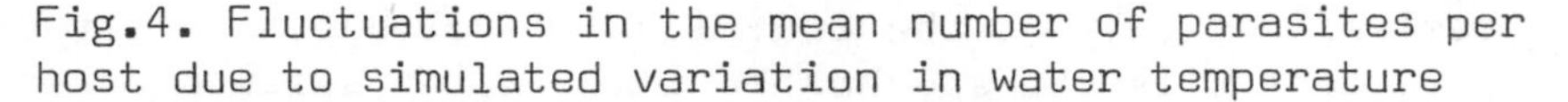
Fig.4. Fluctuations in the mean number of parasites per host due to simulated variation in water temperature

estimate the death rate $\mu(t)$. The results of a typical simulation run are shown in Fig.4 from which it can be seen that natural variation in water temperature does not markedly affect the stability of the cyclic behaviour of the adult parasite population.

At equilibrium in the deterministic model

$$N_t = \lambda(t) \; / \; \mu(t)$$

and since both λ and μ are functions of time there will be a different equilibrium value at each point in time. In the equivalent stochastic model the population will never appear to reach a steady state due to chance fluctuations around the equilibrium. During periods of low numbers of adult parasites these chance fluctuations may result in the population becoming extinct. However, due to the longevity of the larval parasite in the tubificid oligochaete this may not cause the termination of the life cycle. The parasite eggs similarly remain infective for long periods of time. Thus it is possible for both the adult and larval parasite populations to disappear for a short period without the species becoming extinct. This type of behaviour is often caused by seasonal changes in the environment which may either directly influence the parasite populations or alter the behaviour of the host.

A further point to consider is the age structure of the host population. The immigration rates used in the stochastic model of the adult *C. laticeps* were average rates estimated from a stratified sample of the host population. However different age groups of fish have differing feeding habits thus directly influencing the uptake of infective larval parasites. The adult parasites, therefore, may be present in certain age groups of the host population while absent from others (Anderson 1971). These factors illustrate the importance of a detailed biological knowledge of a parasite's life cycle in the assessment of the stability of the host parasite interaction. This knowledge is enhanced by the formulation of a mathematical model to describe the biological processes involved. Ideally, stochastic models can be constructed and parameters estimated for each compartment in a life cycle. The submodels could then be linked together and the behaviour of the complete cycle investigated.

This approach, however, ignores the fact that a host parasite relationship is the interaction between host and parasite, not only at the individual level but also between populations. In the stochastic immigration death process, except for the recognition of heterogeneity in feeding behaviour, the host population has been ignored. Thus, when asking the broad question "How stable is a specific host-parasite interaction?" both populations must be examined simultaneously. A model to describe the interaction should ideally incorporate equations to describe rates of change in the host and parasite populations. Due to the complexities of such an approach the problem is treated in a deterministic manner.

DETERMINISTIC MODELS OF PARASITE LIFE CYCLES

Direct life cycles

The direct life cycle represented diagramatically in Fig.1a provides a convenient starting point for this approach. The information ideally required from the model can be summarized as follows:

(1) Host population size;

(2) Adult parasite population size;

(3) Parasite egg population size;

(4) Distribution of the adult parasites within the host population.

The last point appears difficult to handle at first sight i the framework of a deterministic model. However, the hosts can be regarded as compartments through which the parasites flow due to immigration and death. A host or compartment will contain 0, 1, 2, ... , s parasites where s is a theoretical maximum. This quantity s can be envisaged in two ways. First, it can be thought of as a maximum carrying capacity of the hosts, a limit imposed by the physical size of the microenvironment the parasite occupies. For convenience sake it has to be assumed that the size of a parasite is not determined by the number of parasites present within one host. This it not strictly true of all host parasite interactions as there is evidence that growth rate is density dependent (Read, 1950). The carrying capacity s is also assumed to be the expected or mean value for the population of hosts. Secondly, the parameter s can be considered as a lethal level acting as a limit to the number of parasites a host can harbour. This is a concept first described by Crofton (1971) when he suggested that all parasites will kill their hosts if present in large enough numbers. The physical carrying capacity of the host may never be reached since the lethal level may be smaller than this limit. Each host in a population will vary in its response to the parasite, thus an expected lethal level is assumed for the population. For the purpose of mathematical convenience the lethal level is a more manageable conceptual approach.

The parasites will flow through the compartments or hosts until a value s is reached in any one compartment when both host and the s parasites are removed from the system. It is assumed that in a small interval of time only single transitions of parasites will occur between hosts. The hosts are themselves labelled by the number of parasites they contain, and the separate classes of hosts can be described by the following notation:

Let x_i be the number of hosts containing i parasites where $i = 0, 1, 2, \ldots, s$. Thus

$$\sum_{i=0}^{s} x_i = H, \text{ the total number of hosts}$$

$$\sum_{i=0}^{s} i x_i = P, \text{ the total number of parasites.}$$

Let ω represent the total number of parasite eggs. A set of simultaneous differential equations to describe the rate of change in ω, x_0, x_1, ..., x_s are constructed in the following manner

$$\frac{d\omega}{dt} = \lambda_1 \sum_{i=0}^{s} i x_i - \mu_1 \omega - \lambda_2 \omega \sum_{i=0}^{s} x_i \tag{15}$$

$$\frac{dx_0}{dt} = \lambda_3 \sum_{i=0}^{s} x_i - \mu_3 x_0 - \lambda_2 \omega x_0 + \mu_2 x_1 \quad (16)$$

$$\frac{dx_i}{dt} = \lambda_2 \omega x_{i-1} - \mu_3 x_i - \lambda_2 \omega x_i + \mu_2 (i+1) x_{i+1} - \mu_2 i x_i \quad (17)$$

$$\frac{dx_s}{dt} = \lambda_2 \omega x_{s-1} - \mu_3 x_s - \lambda_2 \omega x_s - \mu_2 s x_s \quad (18)$$

where λ_1 is the birth rate of parasite eggs per adult parasite, μ_1 is the death rate of parasite eggs per egg, λ_2 is the immigration rate of adult parasites per host, μ_2 is the death rate of adult parasites per parasite, λ_3 is the birth rate of hosts per individual, μ_3 is the death rate of hosts per individual. Although unrealistic, all the rates of change are assumed for simplicity to be constant. This type of notational framework was first used in a mathematical description of host parasite interactions by Kostitzin (1934). The non-linearity of the model results from the dependence of the immigration rate of the adult parasite population on the size of the parasite egg population, a factor regarded as an essential biological feature of host parasite systems. Although an analytical solution of the equations has not been found they can be solved by numerica approximation.

A set of non-negative equilibrium values exists for the equations but these states are unstable to small displacements. The stability of the model is greatly enhanced by formulating the death rate μ_3 of the host population as a function of both the number of parasites present in a singl host and the number of hosts in the population, both very realistic biological assumptions. A simple exponential for of the death rate is assumed

$$\mu_3\left(i, \sum_{i=0}^{s} x_i\right) = a \exp \{bi + c \sum_{i=0}^{s} x_i\}$$

where a, b and c are constants. The equations incorporatin these assumptions exhibit damped oscillations towards a set of stable equilibrium points (Fig.5) for a large range of values of the rate of change parameters. The starting

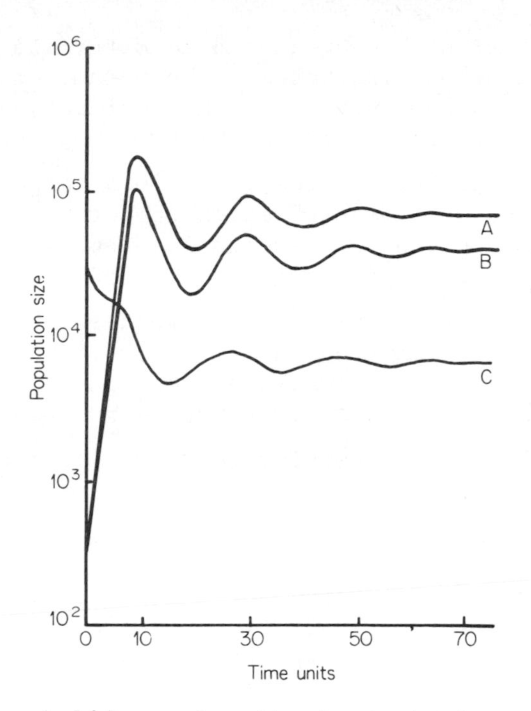

Fig.5. Direct life cycle: the trajectories of host and parasite numbers predicted by the deterministic model. A=parasite egg population, B=adult parasite population, C=host population (Lethal level=15)

values of ω and the vector of x_i's chosen for the numerical solution of the equations influence the time taken to reach a steady state but not the numerical values of the equilibrium points.

A formal stability analysis of the equations as described by May (1973) is inappropriate due to the non-linearity of the system and thus the structural stability of the model (Lewontin, 1969) is difficult to determine theoretically. An assessment of the stability of the model can be achieved by numerical investigation and the following conclusions drawn about the behaviour of the model are based on a large number of numerical analyses using various parameter values. The most important parameters in the model are the lethal level s, the immigration rate of the eggs into the adult parasite population in the host and the birth rate of eggs by the adult parasites. The lethal level has a marked influence on the equilibrium values of ω and the vector of x_i's (Fig.5) the precise relationship depending on the rates of change parameters. This association is not of a simple form and further work is required to determine the precise functional relationships between the parameters. The model is very sensitive to changes in the immigration parameter λ_2. If this rate is too large the host population becomes saturated with parasites, resulting in a large number of host deaths and perhaps extinction of both host and parasite. A high immigration rate can be compensated for by a high death rate μ_2 of the parasites. Under certain circumstances the host population may recover once the parasites have been eliminated due to low host density. If the immigration rate is too small then the parasite population becomes extinct. In this situation ω and $\sum ix_i$ are zero and the host population x_0 settles to an equilibrium point when $\lambda_3 = \mu_3(0,x_0)$.

Alterations in the egg birth rate parameter have a similar effect to that described above for the immigration rate. Too high a rate leads to saturation resulting in host deaths and parasite extinction, while too low a rate results again in failure of the parasites to establish themselves in the host population. In general, although the model is unrealistic due to its simplicity, it is interesting that a wide range of parameter values lead to a stable equilibrium in the host parasite interaction.

It is possible to examine the distribution of the parasite counts within the host, irrespective of the deterministic framework of the model, since the vector of x_i's essentially represents a frequency distribution of the numbers of parasites per host. The mean and variance of this distribution are

$$Y/X \text{ and } (Z - Y^2/X)/(X - 1) \text{ respectively}$$

where

$$X = \sum_{i=0}^{s} x_i, \quad Y = \sum_{i=0}^{s} ix_i \text{ and } Z = \sum_{i=0}^{s} i^2 x_i$$

Observed distributions of parasite counts per host tend to be highly overdispersed and, as shown by the stochastic model of the immigration death process, these may be generated by heterogenity within the host population. Crofton (1971) suggested that overdispersion of parasite counts is essential if stability is to occur in a host parasite system. In the tail of the distribution a few hosts will harbour very large numbers of parasites and it is these few hosts who will suffer most from the infection leading to their eventual death once the lethal level is reached. Thus the removal of a few hosts results in the extermination of a large number of parasites.

Table 2. Direct life cycle: the dependence of the variance:mean ratio on the level s for a fixed set of population parameters

Lethal level, s	Variance/Mean
5	0·383
10	0·954
15	1·244
20	1·343
25	1·488

The deterministic model treats the immigration process as constant for each host in the population and so a factor generating over dispersion is missing. However, numerical analyses indicate that over dispersion as measured by the variance:mean ratio is generated at certain values of s. At the low value of s the distribution is truncated, creating underdispersion, while overdispersion is generated as s increases (Table 2). The values of s which create various patterns of dispersion in the parasite counts are dependent on the population parameters of the model.

Thus, although overdispersion aids in the regulation of the host parasite interaction, it is not essential since stable equilibria occur at low s values when underdispersion is apparent. If s is very large, say infinite for all practical purposes, then the host population will not be regulated at all by the parasites, depending on other processes such as density dependent growth rates. The parasites themselves will be regulated by the balance between immigration and death where the latter may be caused by an immune response of the host.

Indirect life cycles

The framework of the deterministic model can be extended to describe indirect life cycles. Consider the following notation:

ω = number of parasite eggs

y_j = number of intermediate hosts with j larval parasites

x_i = number of final hosts with i adult parasites

where $j = 0, 1, 2, \ldots, r$ and $i = 0, 1, 2, \ldots, s$ and r is the lethal level of the larval parasites and s the lethal level of the adult parasites.

A set of simultaneous differential equations to describe the rates of change in ω, y_j, and x_i for the life cycle of *C. laticeps* (Fig.1b,) are constructed in a similar manner to the direct life cycle case (see Table 3). The added complexity in this model when compared with the direct life cycle case, is due to the distribution of parasites within the intermediate hosts. If a final host with i parasites consumes an intermediate host with j larval parasites, then the final host jumps to the $(i+j)$ class. The cyclic nature of the parameters $\lambda_4(t)$ and $\mu_4(t)$, the immigration and death processes of the adult *C. laticeps*, also adds to the complexity.

The equations are non-linear and thus are solved numerically. For various values of the population parameters the variables exhibit damped oscillations to stable equilibrium points provided λ_4 and μ_4 are constants. However, if these parameters are incorporated as cyclic functions of time, a stable oscillating equilibrium is produced (Fig.6).

The behaviour of the model is very similar to the direct life cycle case which is to be expected due to the identical framework. The model is again sensitive to changes in the immigration and birth parameters λ_1, λ_2 and λ_4 relating to the parasite populations. However, in the life cycle of *C. laticeps*, if the parasites become extinct an equilibrium exists in the predator-prey relationship between final and intermediate host for certain values of the population parameters. This relationship adds complexity to the model and restricts the areas of structural stability. If the final host population, the predator, becomes too large in relation to the intermediate host, the prey population, then the latter becomes extinct and thus the parasite is also eliminated. In natural situations the intermediate host population is usually very large in comparison to the final hosts and also the level of infection of the intermediate hosts with the larval parasites is often low. This type of situation simulated by the model with the corresponding population parameters leads to stability in the host parasite interaction.

A fundamental question often raised in which type of life cycle, direct or indirect, is the more stable? Since indirect life cycles are thought to have evolved from direct cycles (Llewellyn, 1965) what are the advantages of added complexity to surviving in an ever changing environment? Complexity by the inclusion of additional hosts is likely to have evolved in order to enhance the transmission and development of the parasite. However, the added complexity is likely to make the life cycle and populations involved less stable to environmental change, since it is often argued that complexity decreases physical stability of a system (May 1971). An alternative hypothesis is that the increased number of parameters in the complex system may possibly allow greater flexibility in response to change. In the simple direct life cycle a given set of parameter values leads to a stable equilibrium, the removal of one parameter outside this region of parameter stability will often result in a crash of the host or parasite populations which is difficult to reverse by alterations in other parameters. For example, in the direct life cycle in Fig.1a only one infective process exists, between the

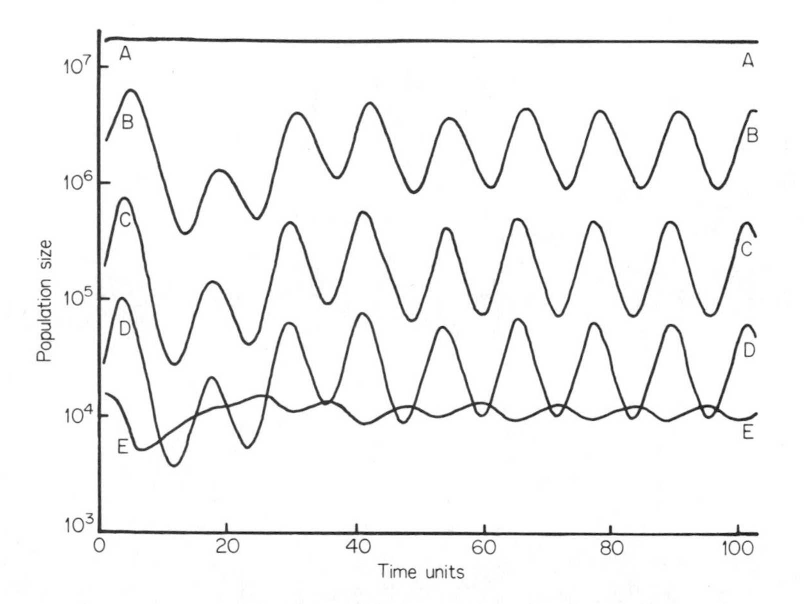

Fig.6. Indirect life cycle: the trajectories of host and parasite numbers predicted by the deterministic model. A=intermediate host population, B=parasite egg population, C=larval parasite population, D=adult parasite population, E=final host population (Lethal level = 30)

Table 3. Simultaneous differential equations for the indirect life cycle. Symbols have the following meanings: λ_1 is the birth rate of parasite eggs per adult parasite; μ_1 is the death rate of parasite eggs per egg; λ_2 is the immigration rate of larval parasites per host; μ_2 is the death rate of larval parasites per individual; λ_3 is the birth rate of intermediate hosts per individual; μ_3 is the death rate of intermediate hosts per individual; λ_4 is the feeding rate of final hosts on intermediate hosts; μ_4 is the death rate of adult parasites per individual; λ_5 is the birth rate of final host per individual; μ_5 is the death rate of final host per individual

$$\frac{d\omega}{dt} = \lambda_1 \sum_{i=1}^{s} i x_i - \mu_1 \omega - \lambda_2 \omega \sum_{j=0}^{r} y_j$$

$$\frac{dy_0}{dt} = \lambda_3 \sum_{j=0}^{r} y_j - \mu_3\left(0, \sum_{j=0}^{r} y_j\right) y_0 - \lambda_2 \omega y_0 + \mu_2 y_1$$

$$- \lambda_4(t)(y_0) \sum_{i=0}^{s} x_i \Big/ \left[\sum_{j=0}^{r} y_j\right]$$

.
.
.

$$\frac{dy_j}{dt} = \lambda_2 \omega y_{j-1} - \mu_3\left(j, \sum_{j=0}^{r} y_j\right) y_j - \mu_2 j y_j + \mu_2 (j+1) y_{j+1}$$

$$- \lambda_4(t)(y_j) \sum_{i=0}^{s} x_i \Big/ \left[\sum_{j=0}^{r} y_j\right] - \lambda_2 \omega y_j$$

.
.
.

$$\frac{dy_r}{dt} = \lambda_2 \omega y_{r-1} - \mu_3\left(r, \sum_{j=0}^{r} y_j\right) y_r - \mu_2 r y_r$$

$$- \lambda_4(t)(y_r) \sum_{i=0}^{s} x_i \Big/ \left[\sum_{j=0}^{r} y_j\right] - \lambda_2 \omega y_r$$

Table 3 - continued

$$\frac{dx_0}{dt} = \lambda_5 \sum_{i=0}^{s} x_i - \mu_5\left(0, \sum_{i=0}^{s} x_i\right) x_0$$

$$- \lambda_4(t)\left[\sum_{j=1}^{r} y_j\right] x_0 / \left[\sum_{j=0}^{r} y_j\right] + \mu_4(t) x_1$$

.

.

.

for $i \leq r$

$$\frac{dx_i}{dt} = \lambda_4(t) \sum_{j=1}^{i} (y_j x_{i-j}) / \left[\sum_{j=0}^{r} y_j\right] - \mu_5\left(i, \sum_{i=0}^{s} x_i\right) x_i$$

$$- \lambda_4(t)\left[\sum_{j=1}^{r} y_j\right] x_i / \left[\sum_{j=0}^{r} y_j\right] + \mu_4(t)(i+1) x_{i+1} - \mu_4(t) i x_i$$

for $i > r$

$$\frac{dx_i}{dt} = \lambda_4(t) \sum_{j=1}^{r} (y_j x_{i-j}) / \left[\sum_{j=0}^{r} y_j\right] - \mu_5\left(i, \sum_{i=0}^{s} x_i\right) x_i$$

$$- \lambda_4(t)\left[\sum_{j=1}^{r} y_j\right] x_i / \left[\sum_{j=0}^{r} y_j\right] + \mu_4(t)(i+1) x_{i+1} - \mu_4(t) i x_i$$

.

.

.

$$\frac{dx_s}{dt} = \lambda_4(t) \sum_{j=1}^{r} (y_j x_{s-j}) / \left[\sum_{j=0}^{r} y_j\right] - \mu_5\left(i, \sum_{i=0}^{s} x_i\right) x_s$$

$$- \lambda_4(t)\left[\sum_{j=1}^{r} y_j\right] x_s / \left[\sum_{j=0}^{r} y_j\right] - \mu_4(t) s x_s$$

free living parasite eggs and the host. If this immigration rate suddenly increases due to a change in the environment, it is only possible to compensate for this by changes in the death rate or indirectly by a decrease in the birth rate of eggs.

In the indirect life cycle (Fig.1b), the alteration of a single parameter can be compensated for by the alteration of one of the many other parameters, allowing the system to recover before extinction occurs of either host or parasite. It is, however, difficult to imagine how such a compensatory mechanism would be triggered in natural host parasite systems.

DISCUSSION

The models discussed in this paper are by their very nature over simplified and thus misrepresent the host-parasite interactions that they profess to mirror.

The stochastic model of the immigration death process is the more precise approach since it deals with a very small and comparatively simple set of biological processes and incorporates a basic heterogeneous process of the host parasite interaction. Variability in transmission processes and previous experiences of infection by the host are essential features of host parasite systems and are exhibited in the overdispersed nature of the distribution of parasite counts per host. In addition, where seasonal periodicity of the parasite population size exists chance elements will be important since during periods of low numbers the probability of extinction of the parasite population will be increased. However, this extinction as mentioned previously may only be a temporary phenomenon due to the longevity of infective stages. It is apparent there fore that stochastic formulations are advisable when modelling host-parasite relationships although such an approach may not always be practical at present due to the inherent complexities of parasite life cycles and a lack of detailed knowledge concerning the parameters of the system.

The deterministic models of the direct and indirect life cycles are over simplified, ignoring such factors as time lags in development, immune response by the host triggered by previous experiences of infection, age structure of both host and parasite populations and spatial heterogeneity. Since these processes are of obvious biological importance, conclusions based on the deterministic models concerning the stability of host-parasite systems must be accepted with caution. However, the framework of the models does describe the essential features of host-parasite interactions and provides information about changes in both host and parasite populations. The models also provide a basis for further elaboration, and it is advisable to investigate fully the properties of simple formulations before attempting to understand the behaviour of more complicated and perhaps more realistic models.

A major disadvantage of the deterministic formulations is the basic non-linear structure of the equations which at present inhibits a fully theoretical investigation of the model. The numerical solution of such a model is time consuming and in some cases it is not practical. For example, if the lethal limit s is very large, say in the region of 1,000, then this requires the solution of over 1,000 non-linear simultaneous differential equations. This type of situation does not arise in the majority of helminth parasite infections but often occurs with protozoan parasites. One possible method of overcoming this problem is to convert the vector of x_i's into a continuous variable which can take any value from 0 to s; thus the summation $\sum x_i$ representing the host population size becomes $\int x(i)di$. This method, however, does raise some problems in connection with the limits of the continuous function, but it does appear to be a promising line for future investigation.

The problems involved in modelling host parasite relationships should not detract from the fact that when dealing with complex population interactions genuine scientific insight into the factors influencing and controlling the dynamics of these populations is unlikely to occur if conducted in largely intuitive and verbal terms. In order to achieve real insight into and control of the biological phenomenon and processes operating, particularly in the case of medically and veterinary important parasite diseases, some more precise, logical framework is required as is provided by the right kind of mathematics.

SUMMARY

Two approaches towards the mathematical description of helminth parasite life cycles are described. The first approach concerns a stochastic model of one compartment of an indirect parasite life cycle, namely the adult parasite population in the definitive host. Heterogeneity in the number of immigrants entering the hosts is incorporated in the model and the resulting distribution of parasite counts per host is derived and shown to be overdispersed. The stability of the model in relation to the influence of the parameters involved and to natural variation in environmental variables is investigated. The second approach concerns the construction of simple deterministic models of both direct and indirect parasite life cycles and provides a framework for the investigation of the interactions between host and parasite populations.

ACKNOWLEDGEMENTS

I am indebted to Professor M.S. Bartlett, F.R.S. for advice and many invaluable discussions. This work was carried out during the tenure of an I.B.M. Research Fellowship.

REFERENCES

Anderson, R.M. (1971). A quantitative ecological study of the helminth parasites of the bream *Abramis brama* (L.) *Unpublished Ph.D. thesis, London University.*

Bailey, N.T.J. (1964). *The elements of stochastic processes with applications to the natural sciences.* London: John Wiley.

Bartlett, M.S. (1955). *An introduction to stochastic processes.* Cambridge: Cambridge University Press.

Conway, G.R., Glass, N.R. and Wilcox, J.C. (1970). Fitting non-linear models to biological data by Marquardt's algorithm. *Ecology*, 51, 503-8.

Crofton, H.D. (1971). A model of host-parasite relationships. *Parasitology*, 63, 343-64.

Kennedy, C.R. (1969). Seasonal incidence and development o the cestode *Caryophyllaeus laticeps* (Pallas 1781) in the River Avon. *Parasitology*, 59, 783.

Kennedy, C.R. (1970). The population biology of helminths in British freshwater fish. *Symp. Br. Soc. Parasit.*, 8, 145-57.

Kostitzin, V.A. (1934). *Symbiose, paracitisme et evolution* Paris: Hermann.

Kulakowskaja, O.P. (1962). Development of Caryophyllaeidae (Cestoda) in an invertebrate host (Russian text). *Zool. Zhur.*, 41, 986-92.

Kulakowskaja, O.P. (1964). Life cycles of Caryophyllaeidae (Cestoda) in the conditions of Western Ukraine (Russian text). *Cslka. Paresit.*, 11, 177-85.

Lewontin, R.C. (1969). The Meaning of Stability. *Diversit and Stability in Ecological Systems.* Brookhaven

Symposium in Biology, No.22, 13-24.
Llewellyn, J. (1965). The Evolution of Parasitic Platyhelminths. *Symp. Br. Soc. Parasit.*, 3, 47-77.
May, R.M. (1971). Stability in model ecosystems. *Proc. Ecol. Soc. Australia*, 6, 18-56.
May, R.M. (1973). On relationships among various types of population models. *Am. Nat.*, 107, 46-57.
Read, C.P. (1950). The 'crowding effect' in tapeworm infections. *J. Parasit.*, 37, 174-8.

Stability in host-parasite systems

DAVID J. BRADLEY

Department of Pathology, University of Oxford

Parasitism is an ecological relationship between at least two populations, parasites and hosts, and the interactions between them are crucial in determining parasite numbers. Much of the theoretical work on population ecology has dealt with two interacting populations and so we may hope that theory and observation may draw closer together in understanding the host-parasite system. However, the superficially simple host-parasite relation is often complicated by the presence of complex life cycles involving additional hosts, and by the occasional presence of free-living stages. After some introductory generalizations about parasites and the modelling of populations, we consider the regulation of parasite populations in conceptual terms and discuss two types of stability in more detail, attempting to combine mathematical modelling, field data and experimental work. These three approaches need to proceed forwards together if we are to understand stability: field data alone lead to confusion, laboratory work in isolation may lead to distortion, and mathematical models alone tend to rise into the clouds.

Parasitic infections are very common in nature. Protozoa and helminths are the main groups that are thought of as animal parasites, though in ecological terms the concepts involved in understanding protozoal infections are largely applicable to other microbes such as bacteria. The parasitic insects are rather better considered as micropredators and are not dealt with here. Several generalizations apply to most parasitic protozoa and helminths:

Ecological Stability
Edited by M.B. Usher and M.H. Williamson.

several or many parasites may infect one host, the parasites have to get from one host to another, they have very high reproductive rates, often the host can acquire some form of resistance to the parasite, and the parasites - if sufficiently numerous - may harm the host.

Yet many host-parasite systems are relatively stable, using the term in a practical sense. There is often less variation from year to year in parasite numbers than would be expected from the variation in factors that might cause variation; catastrophic epizootics are less common than steadier lower levels of infection; and the effects of measures aimed at reducing the level of infection are frequently less dramatic than is hoped for. A clear example of this was seen in Lagos, Nigeria, where attempts at control of malaria transmission succeeded in reducing the anopheline mosquito population to under 2% of its previous level. The effect on human malaria was very slight. Operationally, this is what is meant by stability and it is clearly of both practical importance and theoretical interest. What then is the basis for this relative stability? Clearly it must result from the operation of the processes which affect parasite numbers.

There are two main approaches to the study of this stability. One may attempt to construct a detailed mathematical model of the whole life cycle in terms of sets of differential equations and then examine these mathematically for evidence of stable equilibria. Such an approach is considered by Anderson in this book. It may be thought of as corresponding to the complex ecological models of Watt (1966) and of Paulik and Greenough (1966). The alternative approach, followed here, is to consider rather simple conceptual and mathematical models, closely related to epidemiological and experimental work, in the hope that they may illuminate the main factors responsible for stability. This perhaps corresponds more to the approach of MacArthur and Wilson (1967) and Levins (1968) to the analysis of ecological problems, though in a much less developed form.

THREE WAYS OF REGULATING PARASITE POPULATIONS

Type 1: Populations determined by transmission

Stabilising mechanisms may be approached by considering an unstable population initially. The problem of transmission has particularly interested the classical writers on parasitism. They have emphasised the difficulties of the parasitic life and pointed out that only a small proportion of parasites, often one in a million or less, may effect the transition to a new host. Many parasites have very high reproductive rates - a female *Ascaris* lays over 100,000 eggs daily and a trypanosome may undergo binary fission daily - which compensate for the transmission loss, but with such a large potential for both loss and increase it is clear that great variations in transmission rates may occur. The parasite usually has a shorter life than its host. When the probability of successful transmission varies, so does the total parasite population. Indeed, small changes in transmission, if maintained, may in time lead to large changes in the parasite population though outside the humid tropics many such environmental fluctuations are cyclical.

The transmission process may be affected by environmental factors extrinsic to and independent of the parasite. An example of a situation where transmission is affected by many external environmental factors is given by malaria in an epidemic area, such as the Punjab, where the ambient temperature, humidity, distribution of surface waters, vegetation shading those waters, number of insectivorous birds, may all affect transmission but are scarcely or not at all affected by transmission.

One can imagine parasite populations determined almost wholly by variations in transmission. For this to be so, each inoculum of the parasite into a host must behave independently of the others: thus doubling the input of parasites into the hosts should double the output of parasites. This implies for helminths that crowding levels should not be reached and that they should be self-fertilizing or reproduce asexually. For protozoa it implies that the course of a superinfection should resemble that of a primary infection. These conditions may be approximated at low levels of transmission and when the parasites are of low pathogenicity.

Under these circumstances the variable extrinsic factors, by their effects on the birth and death rates of parasites and on the transmission of the parasite from one host to another, can control parasite numbers, but they only do this in a very imperfect and erratic way. Fluctuations in numbers are the rule, local extinctions are the rule, and life is clearly very uncertain. If there are a lot of mutually independent variables affecting transmission, as they are likely to vary out of phase with each other the transmission rate will tend to be more stable. This has its defects: if the resulting steady rate exceeds one the para-

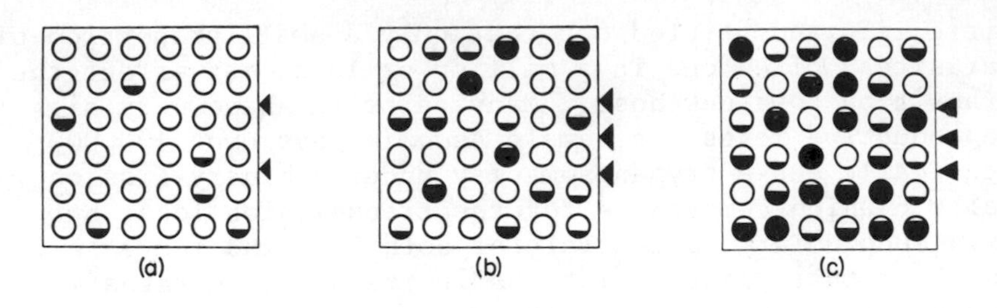

Fig.1. The first 4 Figs. illustrate different types of regulation at 3 transmission levels, low in *a*, moderate in *b* and high in *c*. Each circle represents a host and the black shading its parasite load. The number and size of solid triangles represent intensity of transmission. Fig.1 represents a transmission-determined infection (type 1): the more the transmission the more hosts are infected

site population will steadily increase and at a level below one it will proceed to extinction. Such a host-parasite system, controlled by the extrinsic factors determining transmission, is likely to be unstable and cannot really be said to be regulated.

Mathematical models have been based on a transmission-determined system and these have been of some value in predicting that components of the system have been overlooked (Hairston, 1962) and in describing the form and timing of epidemics, as in the well known malaria model of Macdonald (1957). His emphasis on the occurrence of superinfection has the effect of emphasising transmission at the expense of other regulating mechanisms and though his analysis in

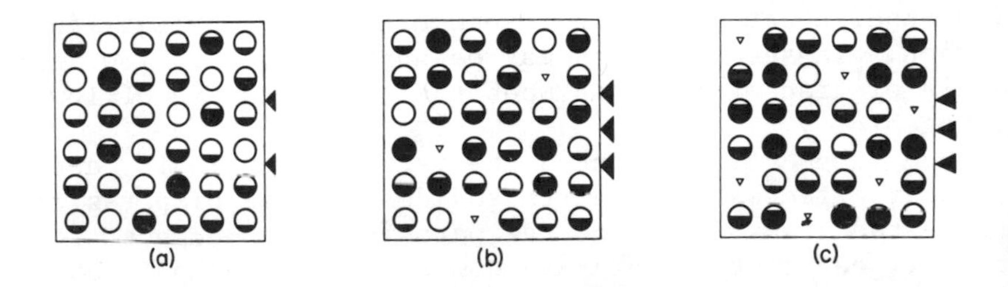

Fig.2. A parasite population limited by death of hosts due to heavy parasitism, limitation at the host population level (type II). A dead host is represented by an inverted triangle. At high transmission levels more hosts die but thereby reduce the numbers of parasites available for transmission

mathematical terms fits best to epidemic states, it was Macdonald's (1951) work that first made clear the relation between endemicity and stability in a parasitic infection.

What I have called transmission-determined infection, where transmission is dependent on the varying environment and in turn determines both the lowest and greatest amount of parasitization observed, is a rather inefficient level of life, an environmentally determined strategy, and is illustrated in Fig.1. It partly corresponds to density-independent regulation (Andrewartha and Birch, 1954) in other animal populations though there are density-dependent aspects of transmission regulation. It characterizes many parasites at the edge of their range, and the parasite population maintained is unstable. It probably also applies to parasites of short lived hosts, where annual replacement of the host stock ensures that parasite populations do not build up over very long periods of time.

Type II: Parasites regulated at the host population level

A parasite which achieves a more efficient level of transmission than that considered above will increase in abundance, and parasite numbers will build up in the host. This does not require phenomenal levels of transmission, for if the basic reproductive rate of the parasite persistently exceeds one the parasite numbers will tend to increase exponentially.

Two simple types of outcome may follow such increase, depending on whether the host mounts an immune response or other check to increase of the parasites. If not, parasites will greatly increase, by addition to the worm load for helminths or by multiplication of the originating infective inoculum for protozoa. Excessive parasite burdens tend to harm, eventually to kill, the hosts and will reduce the number of parasites available for transmission (Fig.2). On the other hand, the host may have the ability to develop resistance to the parasites and kill them (Fig.3). The two types of outcome may be called pathogenesis and immunity respectively, and both may produce some degree of stability of parasite populations.

First, we shall consider pathogenesis. The injurious effects of heavy parasite loads, where those most heavily infected die, tend to control the numbers of hosts as well as parasites. Although the concept of mutual population regulation by host and parasite is highly attractive as a concept, such an isolated system is rare in nature and obvious large-scale host mortality of vertebrates tends to

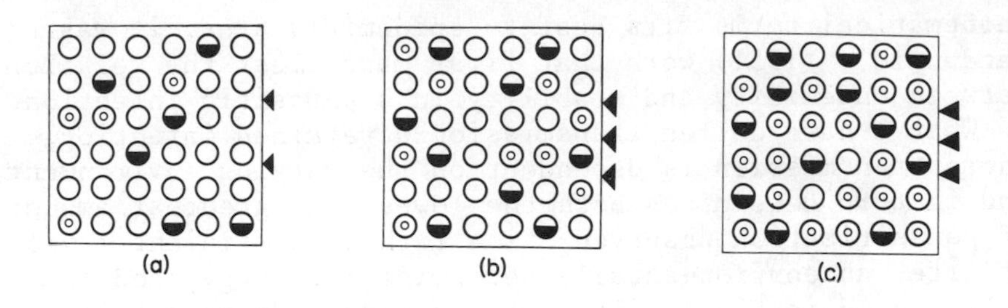

Fig.3. Another parasite population limited by type II mechanisms, by acquired sterile immunity of the hosts. An immune host is represented by the small open circle within it

occur under highly unstable circumstances. But parasitic infections, when endemic, may kill a limited proportion of hosts.

Attention has been drawn (Crofton, 1971b; Bradley, 1972) to the way in which efficiency of regulation by parasite density-dependent host mortality depends on the distribution of parasites among hosts. This effect is seen particularly in those parasites that do not multiply indefinitely within the host.

Parasitic helminths in nature are not usually randomly or evenly distributed amongst their hosts but follow a highly aggregated pattern, so that the majority of the worms are found in a few of the hosts (Li and Hsu, 1951; Bradley, 1965; Thurston, 1968). Much discussion has centred on the precise nature of the distribution (Williams, 1964; Crofton, 1971a) - whether negative binomial, logarithmic or log-normal - but that aggregation of parasites is usual seems clear. The output of variable parasites or their eggs from a host may be at least as

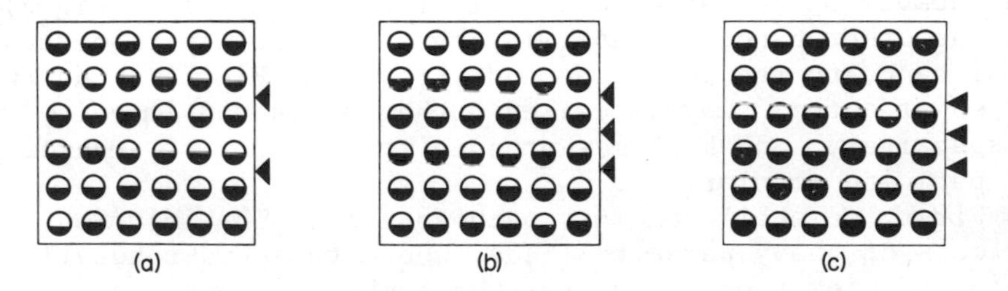

Fig.4. The ultimate in type III regulation by host individuals. Once the individual has acquired a certain level of infection the regulatory mechanism in each host prevents its rising further

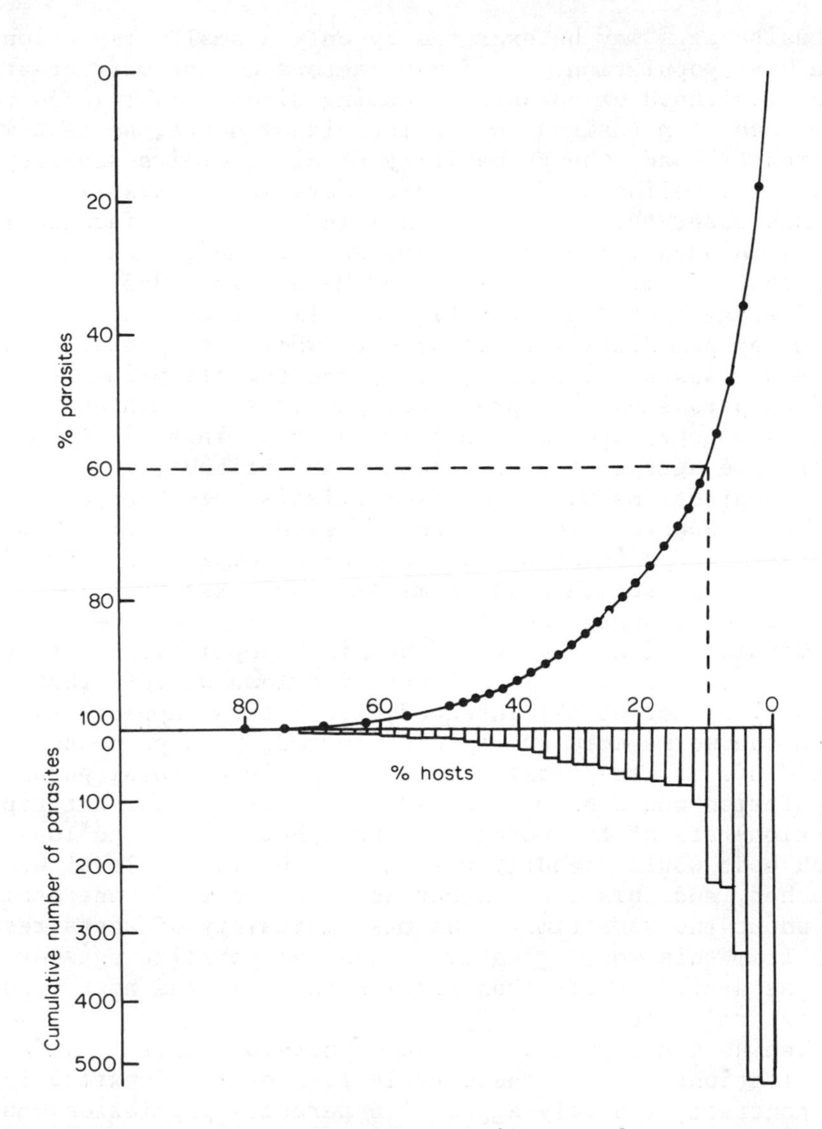

Fig.5. Cumulative host and parasite egg frequency curve for *S. haematobium* in Tanzanian school children. The pattern of egg counts is shown above and from the cumulative graph one may read off the smallest proportion of children (hosts) excreting any given proportion of the eggs (parasites) being disseminated. The dotted lines indicate that 10% of the hosts disseminate 60% of the parasites

aggregated. This is illustrated by a cumulative host and parasite frequency curve (Fig.5) showing that half the parasite eggs may be excreted by only a small proportion of the host population. If other factors do not vary greatly, the likelihood of parasites causing disease is related to the load of parasites on the individual hosts, so that with a greater load, the probability of disease, its severity, and the likelihood of death are increased. This is a common observation in helminthic infections of farm animals, in such infections as hookworm in man, onchocerciasis, and schistosomiasis (Forsyth and Macdonald, 1965).

The aggregated distribution of ill effects and infective forms of parasites will be altered where the parasite has separate sexes (Macdonald, 1965) and the offspring of paired parasites (for practical purposes, helminths) are the pathogenic agents. Such is the case in schistosomiasis where the eggs cause most damage, and probably in those filarial infections where the microfilariae are the pathogenic stages. However, the effects of sex on the distribution of pairs and eggs between hosts are more complex and introduce less aggregation than has sometimes been asserted (Bradley and Gliddon, unpublished).

Considering an aggregated helminth population so that some hosts receive a lethal parasite load whilst others are lightly or not at all infected, it at first appears that, were the worms more evenly distributed, perhaps no host would die of its parasite load and a higher total parasite population would be maintained. However, if the net reproduction rate of the worms remained above unity the load on each host would steadily rise until the lethal level was reached, and this would occur for many hosts in one habitat at about the same time. The mass mortality of hosts resulting from this would greatly reduce the parasite eggs or larvae available for transmission and also the host might be exterminated.

The helminth population would therefore undergo wide oscillations with a considerable risk of local extinction. By contrast, a highly aggregated parasite population would result in a large part of it being confined to a few of the available hosts. With rising worm numbers those most heavily infected would suffer and die, with a resulting marked fall in reproductive bodies of the parasite available for transmission. Considerable regulation of the helminth population would therefore follow from the death of few hosts. The regulating mechanism would be highly damped, allowing the maintenance of a relatively stable population of both worms and hosts. A consequence of this pattern is that most hosts will appear relatively underexploited by the parasite, and the usually observed pattern of many infected, some sick and few dying is produced.

Thus a stable host-parasite system is conceivable on the basis of an aggregated distribution of parasites and host mortality dependent on parasite density. Does such a system exist in nature, and does it stand up to more rigorous mathematical study?

There is ample evidence from the field that parasites are aggregated and that very heavily infected hosts are more likely to die. No clear evidence exists that the scale of these phenomena is responsible for regulation of a natural host-parasite system. Crofton (1971b) has elegantly shown for an acanthocephalan parasite of *Gammarus* that, if a negative binomial distribution be generated by the aggregated infective process, then the death of heavily infected hosts is implied by the observed distribution of parasites among the *Gammarus*. The lethal levels of parasites per host in his system are, however, lower than those observed in direct laboratory studies.

Simple mathematical models were used by Crofton (1971b) and by us (Gliddon and Bradley, unpublished) to examine the degree of stability attained at differing lethal parasite loads and levels of aggregation. Crofton, using low lethal levels, obtained good stability. We used rather higher lethal levels and found that except with extreme aggregation the cost of stability in terms of host deaths was very high. About a quarter of hosts would die in each cycle in one example and this appears unlikely to persist in view of the extreme selection pressure that would be exerted on the host. Stability may be produced in this way for helminth parasites of small invertebrates, but appears unlikely for vertebrate hosts.

Secondly, we shall consider immunity. The vertebrates have evolved the ability to respond to foreign antigens. The host may mount a successful immune response and all the parasites be destroyed by it, leaving the host parasite-free and immune. This is the classical response to many viral and bacterial infections. In these situations the individual host can scarcely be said to regulate the parasites, it simply exterminates them locally within itself. However, the result of this process occurring many times over in a host population is to regulate the parasite population considered as a whole (Fig.3). The precise pattern of regulation produced by this immune system or by the death of heavily infected hosts depends on the size of the host population involved and on the details of transmission.

In the short run a parasitic infection that produces complete immunity does not much differ from one that kills its host in the effect on the parasite population, though the depletion of breeding hosts will ultimately have a very different effect. This is the classical infective disease situation analysed mathematically by Kermack and McKendrick (1927) and many others since then (see Bailey, 1957) where differential equations for rates of birth, infection and recovery to immunity are set up for the hosts, and the parasites within any host are only considered together, to give a 'case' of the infection. The analyses usually give an epidemic picture though more complicated mathematical models may give a relatively stable or endemic pattern. The most careful studies, backed by field data, have been on measles, a virus infection, and the minimum host population size for prolonged virus maintenance is well over 100,000 people.

Both these types of regulation at the host population level, aggregation-mortality and complete immunity, are highly dependent on the spatial structure of the host population and of the transmission of infection; upon the 'geometry of living'. The conditions for stability are therefore restrictive.

The protozoa differ from most pathogenic bacteria in having a lower maximal reproductive rate within the host. The staphylococci and enterobacteria may undergo binary fission more than once an hour whereas even the most virulent trypanosomes divide little more than once daily and the maximal rate attained by *Plasmodium falciparum* corresponds to doubling every 10 hours. Since the infective protozoal inoculum is inevitably quite small, the time taken for parasite numbers to rise in a vertebrate host to a transmissible level will be of the same order as, or longer than, the time required to mount an immune response. Thus the 'smash and grab' life strategy of the cholera vibrio or measles virus is not feasible. Those microbes depend on reaching a transmissible level of infection in th host before being immunologically destroyed. The animal parasite has to cope with its host's immune response and, i it is to succeed, to find a way partly to avoid its effects. The immune response often provides a ready-made host mechanism to limit parasite reproduction and has been exploited in the third type of host-parasite system.

Type III: Parasites regulated by host individuals

We have suggested that, though regulation of a parasite pop ulation at the host population level can be achieved by varied mechanisms and attain some precision, yet an efficient mechanism will be very dependent on a particular host community's size and structure. It is therefore a precarious regulating mechanism in the real world.

An altogether more stable situation results if transmission is continually well above the minimum for persistence of the parasite, and numbers are regulated by the individual hosts. If we consider a helminth whose usual transmission rate is 100 worms inoculated per host and the host has a means of preventing the worm load exceeding 10, then over a tenfold fall or indefinite rise in transmission the parasite population will be perfectly regulated as illustrated in Fig.4. This is the ultimate in regulation: highly efficient transmission combined with 'premunition' or some similar process of parasite population regulation by each host. It is seen to a varying degree in very many parasitic systems. Since hosts are not readily able to measure the exact size of their parasite load, mechanisms based on maintaining the load absolutely steady are not usual but a variety of tactics have been evolved by host and parasite that produce what approximates to a similar result.

In the protozoa, the infective inoculum is very small relative to the lethal parasite burden for most hosts. The parasites multiply within the host. It follows that regulation can be achieved by inhibition of parasite reproduction once a moderate parasite level has been reached. Such inhibition would also prevent superinfections contributing appreciably to the parasite load. This reproductive inhibition occurs in *Trypanosoma lewisi* (Taliaferro, 1924; D'Alesandro, 1962) and in *Leishmania donovani* in mice (Bradley, 1971) to an imperfect degree. In other hosts local infection persistence may be the price of an immunity to superinfection or to generalized spread of the parasite. In *T. lewisi*, a natural infection of rodents, temporary stability is achieved in that the rat produces an antibody early in the infection which inhibits parasite reproduction so that a steady high trypanosome population persists until other lytic antibodies are formed.

Trypanosomiasis of man (sleeping sickness) and cattle (nagana) in Africa due to trypanosomes of the *brucei* subgroup are flagellate protozoal infections transmitted by tsetse flies. If the parasites continued to multiply at their initial rate of fission, the host would be dead in very few weeks. However, although untreated infections are usually eventually lethal, the infected host may often survive for a year or more so that some powerful process stabilizing the infection must be present. When the detailed course of parasitaemia is followed, parasite numbers rise for about two weeks and then fall rather abruptly. Before trypanosomes quite vanish from the blood they begin to increase again, and numbers rise for another two weeks before a second crisis. The parasite numbers oscillate in this way for the remainder of the infection. It has been shown that the fall is due to lytic antibody production (Gray, 1965) but that trypanosomes from subsequent peaks are insusceptible to the lytic effect of the antecedent antisera. Antigenic variation of the trypanosome surface has occurred and is regularly induced (possibly by antibody) for up to 30 or more cycles. Thus a form of stability results from alternating host responses and parasite antigenic variation. Similar phenomena occur in malaria and possibly some helminth infections.

The mechanisms that have been evolved to stabilise the parasite load resulting from the initial infection in these protozoa will also handle superinfecting inocula: in *T. brucei* the lytic antibody to the initial antigenic strain will destroy superinfections as well, and the antibody inhibiting reproduction by *T. lewisi* will ensure that superinfections remain of a trivial size.

Most helminths of vertebrates differ from protozoa in that the adult worms do not complete their life cycle within the vertebrate host. Therefore more worms are acquired only by superinfection and not by multiplication of those already present in the host. Some form of immunity to superinfection would seem the obvious way to achieve stability, and in one system there is evidence that this may be the mechanism involved. It was shown by Smithers *et al.* (1969) and Smithers (1972), that in Rhesus monkeys infected by *Schistosoma mansoni* the worms take up the host antigens into their surfaces as they mature. Antibody, which can inhibit growth or even kill very young schistosomes, is induced by adult worms. The initiating and inducing infection is preserved by the host antigen from the effects of antibody. Thus challenge infections are unsuccessful whilst the initial infection persists. This is potentially a powerful stabilizing mechanism for the schistosome population. There is some epidemiological evidence suggesting that such a mechanism may be operating in some human populations infected with *Schistosoma haematobium*, the cause of urinary bilharziasis (McCullough and Bradley, 1973; Bradley and McCullough, 1973). Observations of schistosomal egg output over several years in several individual infected children showed a marked relative stability of egg excretion, suggesting some form of regulation of the parasite population by the hosts. The stable ranking of infected children by egg output rather favoured the existence of concomitant immunity with early infections persisting and behaving similarly in the children while superinfections were prevented, although other evidence showed that transmission was

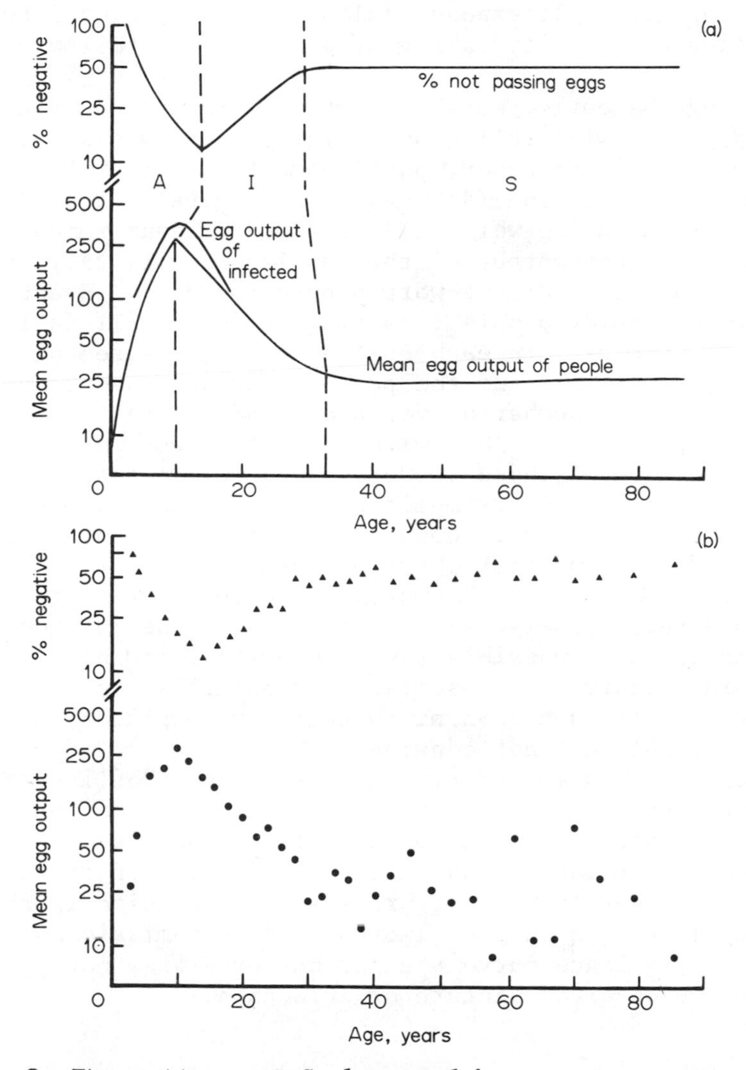

Fig.6. The pattern of *S. haematobium* egg output observed in an endemically infected human population in Tanzania. The three phases of parasite acquisition (*A*), concomitant immunity (*I*) and steady state (*S*) are seen in diagrammatic form in (a), and (b) shows the actual data. Redrawn from Bradley and McCullough (1973)

continuing in the area. Starting from this assumption, and adding the further premises of constant exposure to transmission (this is untrue in reality of course, but no better data were available and the epidemiologically more likely situation of fairly constant exposure in childhood with a gradual decline during adult life would not affect the reasoning greatly), exponential death of worms, and the acquisition of immunity taking some years, the following picture results.

During the early years of life the proportion of uninfected people will fall exponentially, giving a straight line on log-linear paper, until almost all are infected. This is the infection (*A*) phase (see Fig.6a). There follows a period during which all have concomitant immunity (*I*) and the mean egg output of the population will fall exponentially as the resident worm population dies. Eventually the proportion of people remaining infected will fall as the last worm pair in each host dies. The degree of immunity then falls and the person returns to a susceptible state. After a number of years a steady state for the older people results (*S*), with loss of infection balanced by reinfection of those rendered susceptible again. This is clearly a very oversimplified concept and assumes several variables to be constant. This being so, it is surprising to find that observed data from an area of highly endemic *Schistosoma haematobium* infection fit these predictions very closely, as seen in Fig.6. The inference is that it is quite possible that concomitant immunity may occur in urinary schistosomiasis in nature as well as being an experimental phenomenon. Alternative explanations are quite possible but not considered here.

Such a mechanism greatly increases the stability of a parasite population. If the induction of concomitant immunity is faster with a heavy worm load, that is, if it is load-dependent rather than time-dependent, then it will provide very great stability, provided transmission is above a minimum level. A further increase in transmission will make the age-prevalence curve steeper but otherwise not greatly change the observed picture of infection.

CONCLUSION

This short review of a potentially immense subject has shown how very limited our knowledge is - though some understanding of stability in host-parasite systems is essential for those who would control infective diseases of man or domestic stock or who wish to spread pathogens of injurious animals. The failure of massive attempts at malaria eradication in areas of stable malaria, notably in Africa, has shown the practical importance of this subject. Wherever there is a redundancy of transmission above that required for mere persistence of the parasite, various stabilising mechanisms evolve. I have attempted to show that there are a few basic strategies that may readily be followed and a greater range of immunological tactics that may be used to implement them. Various fairly simple models exist which illustrate these forms of stability; some are conceptual and some mathematical. These aid understanding: none is so developed as to be a complete model of happenings in the field. There is therefore much room for further work both at the ecological and epidemiological level and in seeking mechanisms. Wherever a mechanism has been carefully sought it has disclosed phenomena of great basic biological interest such as antigenic disguise and antigenic variation. Stability in host-parasite systems thus involves population, organism and molecular phenomena and only by study of these levels together can valid understanding be reached.

REFERENCES

Andrewartha, H.G. and Birch, L.C. (1954). *The Distribution and Abundance of Animals.* Chicago: Chicago University Press.

Bailey, N.T.J. (1957). *The mathematical theory of epidemics.* London: Griffin.

Bradley, D.J. (1965). A simple method of representing the distribution and abundance of endemic helminths. *Ann. trop. Med. Parasit.*, 59, 355-64.

Bradley, D.J. (1971). Inhibition of *Leishmania donovani* reproduction during chronic infections in mice. *Trans. R. Soc. trop. Med. Hyg.*, 65, 17-8.

Bradley, D.J. (1972). Regulation of parasite populations. A general theory of the epidemiology and control of parasitic infections. *Trans. R. Soc. trop. Med. Hyg.*, 66, 697-708.

Bradley, D.J. and McCullough, F.S. (1973). An analysis of the epidemiology of endemic *Schistosoma haematobium. Trans. R. Soc. trop. Med. Hyg.*, 67, 491-500.

Crofton, H.D. (1971a). A quantitative approach to parasitism. *Parasitology*, 62, 179-93.

Crofton, H.D. (1971b). A model of host-parasite relation-

ships. *Parasitology*, 63, 343-64.
D'Alesandro, P.A. (1962). *In vitro* studies of ablastin, the reproduction-inhibition antibody to *Trypanosoma lewisi*. *J. Protozool.*, 9, 351-8.
Forsyth, D.M. and Macdonald, G. (1965). Urological complications of endemic schistosomiasis in school-children. *Trans. R. Soc. trop. Med. Hyg.*, 59, 171-8.
Gray, A.R. (1965). Antigenic variation in clones of *Trypanosoma brucei*. *Ann. trop. Med. Parasit.*, 59, 27-36.
Hairston, N.G. (1962). Population ecology and epidemiological problems. *Bilharziasis* (Ed. G.E.W. Wolstenholme and M. O'Connor), pp.36-62. London: Churchill.
Kermack, W.D. and McKendrick, A.E. (1927). A contribution to the mathematical theory of epidemics. *Proc. R. Soc.* (A), 115, 700-21.
Levins, R. (1968). *Evolution in Changing Environments*. Princeton: Princeton University Press.
Li, S.Y. and Hsu, H.F. (1951). On the frequency distribution of helminths in their naturally infected hosts. *J. Parasitol.*, 37, 32-41.
MacArthur, R.H. and Wilson, E.O. (1967). *The Theory of Island Biogeography*. Princeton: Princeton University Press.
McCullough, F.S. and Bradley, D.J. (1973). Variation and stability in *Schistosoma haematobium* egg counts. *Trans. R. Soc. trop. Med. Hyg.*, 67, 475-90.
Macdonald, G. (1951). Community aspects of immunity to malaria. *Br. med. Bull.*, 8, 33-6.
Macdonald, G. (1957). *The Epidemiology and Control of Malaria*. London: Oxford University Press.
Macdonald, G. (1965). The dynamics of helminth infections, with special reference to schistosomes. *Trans. R. Soc. trop. Med. Hyg.*, 59, 489-506.
Paulik, G.J. and Greenough, J.W. (1966). Management analysis for a salmon resource system. *Systems analysis in Ecology* (Ed. K.E.F. Watt), pp.215-52. New York: Academic Press.
Smithers, S.R. (1972). Recent advances in the immunology of schistosomiasis. *Br. med. Bull.*, 28, 49-54.
Smithers, S.R., Terry, R.J. and Hockley, D.J. (1969). Host antigens in schistosomiasis. *Proc. R. Soc.* (B), 171, 183-94.
Taliaferro, W.H. (1924). A reaction product in infections with *Trypanosoma lewisi* which inhibits the reproduction of the trypanosomes. *J. exp. Med.*, 39, 171-90.
Thurston, J.P. (1968). The frequency distribution of

Oculotrema hippopotami (Monogenea: Polystomatidae) on *Hippopotamus amphibius*. *J. Zool., Lond.*, 154, 481-5.
Watt, K.E.F. (1966). *Systems Analysis in Ecology*. New York: Academic Press.
Williams, C.B. (1964). *Patterns in the Balance of Nature*. London: Academic Press.

Assessment of the effectiveness of control techniques for liver fluke infection

G. GETTINBY

Department of Mathematics, New University of Ulster

INTRODUCTION

Liver fluke cause damage in sheep and cattle in Ireland estimated to cost the farming industry well over £10,000,000 a year. In Britain the damage to sheep and cattle may be as much as £50,000,000 per year. Ideally one would like to eradicate this parasite; certainly it must be controlled.

The fluke inside its mammalian host is the adult stage of the life cycle of the parasite *Fasciola hepatica* L. While in the host, the fluke produces very large numbers of eggs which pass to the grass via the animal's faeces. There the eggs hatch into a larval stage known as miracidia, which enter a secondary host, the snail *Lymnaea truncatula*. Within the snail the parasite passes through other larval stages involving asexual multiplication. The final intra-snail larval form is shed from the snail back on to the grass; after a further metamorphosis the parasites are in the form of cysts known as metacercariae which are ingested by grazing mammals where they penetrate the liver and mature into adult fluke form, thus completing the life cycle.

One fluke is capable of producing upwards of 10,000 eggs per day for periods measured in years. The degree of infection of the mammalian hosts is stable if on average only one of this astronomical number of eggs survives all the stages of the helminth life-cycle to adulthood. To effect complete eradication of the disease, man would have

Ecological Stability
edited by M.B. Usher and M.H. Williamson.

to kill the sole surviving helminth from the many millio produced by its parent, nature having successfully elimi ated the rest.

Control can be attempted in three ways, each hopefull breaking the life cycle: treatment of mammals, treatment of snails, segregation of the two hosts. The first can done by orally administering flukicides to the sheep and cattle, designed to cure the animal and kill the flukes. The second by treating the pasture with molluscicides designed to remove the snail. The third may be achieved either by fencing off snail habitats to prevent the mamm dropping its faeces where the eggs can find snails or to prevent the mammals grazing on metacercariae-infected grass, or controlling the quality of the pasture e.g. by improved drainage, so as to render it unsuitable as snai habitat. All these controls cost money and any effectiv control policy must balance the cost against the gains obtained for decreased damage to the mammal stock.

Both within the mammal host and the snail there is multiplication of numbers; moreover the multiplication within the mammal is several million times greater than the multiplication within the snail. Between hosts ther is heavy natural mortality of eggs, miracidia and metace cariae, and frequently mortality within hosts due to ove infection. In areas of County Mayo where we have studie the problem, the disease is endemic among sheep, the pro ortion of the sheep population infected at any time bein at least 4-5 times greater than the proportion of the sn population carrying the parasite. The parasite would therefore appear to be most vulnerable to attack at the snail stage of its life-cycle.

Models are proposed here to describe the events leadi to absorption of miracidia in snails. First we give a model of encounters defined as the successful location o snails by miracidia. An absorption model follows, in wh we attempt to relate the successful penetration of the snails by the miracidia which have encountered them. By killing flukes, flukicides reduce the egg output and consequently the number of miracidia. Thus we have to consider absorption and encounter rates as functions of miracidia density. Molluscicides on the other hand remo snails and so ultimately reduce the metacercariae densit which controls the number of adult flukes in the next cycle. We therefore have to determine absorption and encounter rates as functions of snail density. This we by simulation techniques.

ENCOUNTER MODEL

It is well known that there is a chemotactic reaction by which the miracidia become aware of the proximity of snails. This reaction in the case of *F. hepatica* is effective up to some 15 cms.

Let A be the area of the region which is snail habitat. If $P_1(x,y)$ and $P_2(x,y)$ are the positions in this region of a miracidium and a snail respectively, x and y being rectangular co-ordinates, then if $|P_1(x,y) - P_2(x,y)| \leq 15$ the parasite 'encounters' the snail.

Miracidia disperse from cohorts of eggs deposited in clusters randomly in the region. The snails may also be assumed to be distributed randomly in the habitat. If polar co-ordinates are used to measure the position of miracidia relative to their cohort's origin, the parasite spread may be represented by the joint probability density function

$$p(r,\theta)dr\ d\theta = \lambda e^{-\lambda r}\ dr \cdot \frac{1}{2\pi}\ d\theta$$

This assumes:

(1) the distribution is radially symmetric i.e. is uniformly distributed in θ over $(0, 2\pi)$,

(2) the distance from the cohort's origin is exponentially distributed with mean $1/\lambda$, and

(3) r and θ are independently distributed.

Biological evidence suggests that, typically, the mean distance $\bar{d}$ travelled by newly emerged miracidia during their short period of survival in this form is 50 cm. Thus $1/\bar{d} = 0{\cdot}02$ gives an estimate of the dispersal parameter λ.

In order to simulate encounters we first consider a habitat containing s snails and one cohort of m miracidia which originated at $C_0(x,y)$. Such cohort origins can be taken as randomly distributed within the habitat and can be obtained randomly by selecting $x = u_1$, $y = u_2$ where u_1 and u_2 are uniform random variables in the range $\left[0, A^{0\cdot5}\right]$. The position of the jth snail $(j = 1,2,\ldots,s)$ within the region is $P_2(j)$ and can be obtained randomly by selecting $x = u_3$, $y = u_4$ where u_3 and u_4 are again uniform random variables in the range $\left[0, A^{0\cdot5}\right]$. The position of the ith parasite $(i = 1,2,\ldots,m)$ relative to its origin C_0 can be randomly selected by sampling r and θ. Using standard sampling theory (Tocher, 1963) and transforming from polar to rectangular co-ordinates, the position of the ith parasite relative to C_0 is (ε, μ) where ε and μ are distributed so that

$$\varepsilon = -\frac{1}{\lambda} \ln (1-u_5) \cos 2\pi u_6$$
$$\mu = -\frac{1}{\lambda} \ln (1-u_5) \sin 2\pi u_6$$

where u_5 and u_6 are uniformly distributed variables in the range $[0,1]$. Thus the position of the ith parasite can be obtained by sampling $(u_1\ u_2)$, $(u_5\ u_6)$ and adding the described transforms to give a point $P_1(i)$. If the distance $P_1(i)P_2(j) \leq 15$ an encounter occurs.

ABSORPTION MODEL

For this we establish a stochastic model assuming an absorption rate β so that if n is the number of miracidia which encounter a particular snail, then the usual formulation gives

Prob $\{r$ miracidia are absorbed at time $t + \delta t\}$
= Prob $\{(r-1)$ miracidia absorbed at time t and one new absorption occurs in time interval $(t, t + \delta t)\}$
+ Prob $\{r$ are absorbed at time t and no new absorptions occur in time $(t, t + \delta t)\}$

leading to the differential-difference equation

$$\frac{dp_r(t)}{dt} = \beta\{(n-r+1)p_{r-1}(t) - (n-r)p_r(t)\}$$

where $p_r(t)$ is the probability that r are absorbed at time t, with boundary conditions

$$p_r(0) = \begin{cases} 1 & r=0 \\ 0 & r\geq 1 \end{cases}$$

The solution is

$$p_r(t) = \frac{n!}{(n-r)!} \sum_{j=1}^{r+1} \frac{e^{-\beta(n-j+1)t}}{(-1)^{r-j+1} \; (r-j+1)! \; (j-1)!}$$

$$= \frac{n!}{(n-r)! \; r!} \; e^{-\beta(n-r)t} (1-e^{-\beta t})^r \qquad (1)$$

which is a binomial distribution with mean $n(1-e^{-\beta t})$.

This model is imperfect in so far as it ignores the possibility that n may change with time. However, after hatching, the miracidia die within a day unless they are absorbed. The rates of movement of both snail and miracidia are small so that n is unlikely to vary much during the time considered, and is taken as constant in the model. For the history of any cohort of miracidia the probability that in an encounter of n miracidia with a snail r absorptions occur is $p_r(1)$, where the unit of time is the life-span of the miracidia.

The only results known to us, those of Roberts (1950), relate to the absorption of 5 miracidia by single snails in 98 trials under controlled laboratory conditions. The experimental data gives an estimated value of the absorption parameter $\beta = 0{\cdot}439$.

To simulate absorptions from n parasites encountering a snail we first form the cumulative distribution function

$$F(r) = \sum_{j=0}^{r} p_j \; (1)$$

where p_j is the binomial probability defined by Equation 1. Then, generating a random variable u_7 uniformly distributed in the interval [0,1], the value of r such that

$$F(r) \leq u_7 < F(r+1)$$

is the number of absorptions resulting from this encounter.

SIMULATION RESULTS

One of the problems of simulation is to decide how many trials are to be performed to obtain satisfactory results. A series of simulation experiments with models described above showed that the number required was surprisingly small.

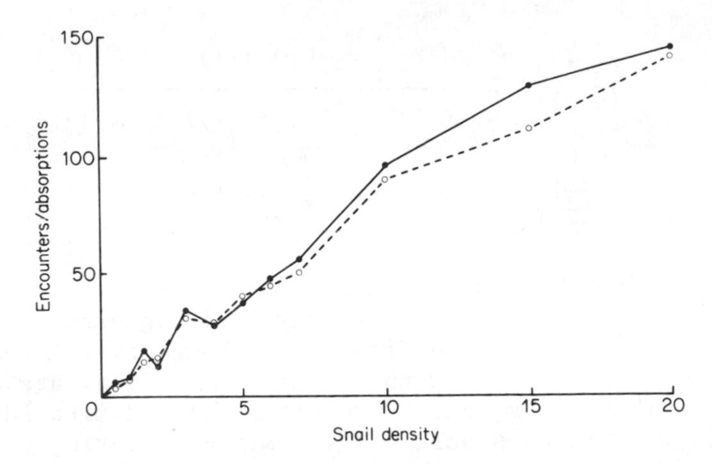

Fig.1. Encounters (dotted line) and absorptions (continuous line) plotted as a function of snail density

The means and variances of encounters and absorptions were stable after 16 trials at most, and further trials did little to improve the stability of the results.

Two sets of results are shown in graphical form. In the first the snail density was varied from 0·5 snails m^{-2} to 20 snails m^{-2} for a fixed miracidia density of 1,000 per cohort. A linear relationship was obtained between snail density and encounters of absorptions as shown in Fig.1. In the second the miracidia density was varied from 50 to 10,000 per cohort for a fixed snail density of 2 snails m^{-2}. The 'logistic' type curve as in Fig.2 was obtained. Little increase in the encounter and absorption rates occurs for parasite populations above the relatively small level of about 3,000 per cohort.

SENSITIVITY ANALYSIS

The effect of varying the parameters β and λ was studied by replicating the simulations with random choice of the parameters in the range $0{\cdot}01 \leq \gamma \leq 0{\cdot}03$ and $0{\cdot}22 \leq \beta \leq 0{\cdot}66$. The variation in λ is equivalent to sampling the mean distance of the parasite from its origin over the range 33·3 c to 100 cm. The distances travelled by miracidia do in fact vary from day to day according to local weather conditions; for example, when the ground is wet they travel further.

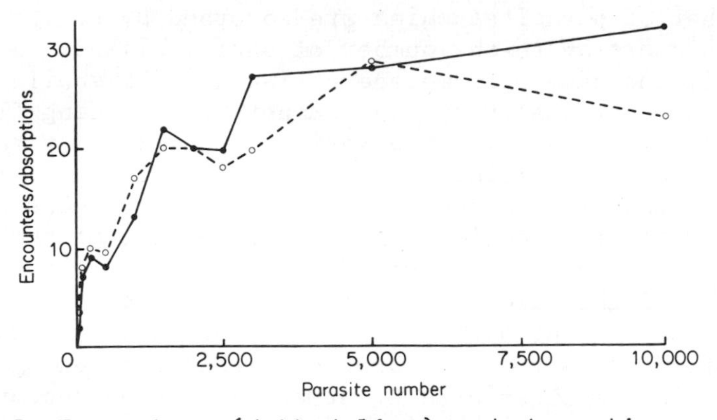

Fig.2. Encounters (dotted line) and absorptions (continuous line) plotted as a function of the number of parasites

The range of daily distances covered in these simulation experiments are considered quite typical for conditions in the West of Ireland and other areas where liver fluke is endemic.

Table 1. Some typical mean absorptions found from the sensitivity analyses

Snail Density (snails m^{-2})	2	15	2
Parasites per cohort	1000	1000	3000
Fixed γ,β	13•2 + 5•3	129•3 + 30•6	27•4 + 9•3
Varied γ,β	11•6 + 6•0	128•3 + 35•5	32•2 + 8•3

Some typical results from the sensitivity analysis are shown in Table 1. The analyses indicate that the model is not over sensitive to changes in values of β and λ. In all cases, however wide the variations of parameters β and λ within the ranges examined, we found that the basic result remained unchanged. Encounters and absorptions depend linearly on snail densities, but only logistically with a rapid approach to an upper limit as parasite densities increase.

CONCLUSIONS

Certain consequences seem to follow the above experimental findings. First, the effect of molluscicides is to reduce the number of parasites which are absorbed by snails in direct proportion to the number of snails killed. Subsequently the number of metacercariae shed by snails is reduced proportionally and the number of impending fluke infections in the mammal population is accordingly reduced. However the molluscicide on its own does nothing to improve the condition of infected mammals - it only reduces the rate of further infection. Moreover, because of the remarkable reproductive powers of the snail population the reduction of the rate of mammal infection lasts only as long as the current parasite life cycle. There is no evidence of accumulative amelioration within succeeding cycles. Thus molluscicide treatment has to be repeated regularly in order to obtain a controlled reduction in the mammal infection rate. There appears to be little likelihood of molluscicide treatment by itself ever achieving eradication of the disease.

Secondly, the effect of flukicides (correctly administered) is to achieve an immediate improvement in the quality of the mammal host by killing its fluke. However, for the total mammal population this improvement is temporary, for, unless the flukicide is successful in killing every fluke in the mammal some eggs continue to be laid. In view of the large numbers of potential miracidia developing from each metacercariae ingested by a mammal host, the smallest margin of failure by the flukicide results in significantly large numbers of eggs being laid. Thereafter, the 'logistic' type relationship between miracidia densities and absorptions ensures that the administration of flukicides does almost nothing to reduce the size of the next generation of metacercariae.

Finally, we quote Macdonald (1965) writing on the dynamics of helminth infections in humans, who said, 'Heavy contamination of the water, though it makes introduction of the infection easier, appears to have virtually no effect on the level of endemicity attained, which is almost exclusively dependent on the number of snails, ...' Our findings, based on the absorption/encounter models, are precisely the same. It is the snail population which has t be controlled. Macdonald also said, 'Treatment carried out to an extent which does not approach eradication of the infection is, therefore, of great value to the recipients but is of little value to others'. This we believe is particularly true of the use of flukicides on domestic animals.

ACKNOWLEDGEMENTS

The author is indebted to Professor A. Young, New University of Ulster, for his guidance and encouragement. He would also like to thank M.J. Hope-Cawdery, Agricultural Institute, Co.Mayo; Professor J.N.R. Grainger, Trinity College, Dublin; Dr. C.B. Ollerenshaw, Ministry of Agricultural and Fisheries, Weybridge; and Miss S.I. McClean, New University of Ulster for their interest and valuable counsel. The work was supported by a research studentship from the Ministry of Education, Northern Ireland.

SUMMARY

The economic implications in the control of liver fluke are far-reaching. Yearly losses in Ireland and the U.K. total well over £60,000,000 and may justify a national control policy. In this paper a mathematical analysis of the interaction of the parasite and snail host populations provides information on the level of infection. Simulation of the models obtained in the analysis establishes quantitative relationships that enable the success of control techniques for eradicating the liver fluke to be measured.

REFERENCES

Macdonald, G. (1965). The dynamics of helminth infections, with special references to schistosomes. *Trans. R. Soc. trop. Med. Hyd.*, 59, 489-506.

Roberts, E.W. (1950). Studies on the life cycle of *Fasciola hepatica* and of its snail host, *Lymnaea truncatula* in the field and under controlled conditions in the laboratory. *Ann. trop. Med. Parasit.*, 44, 187-206.

Tocher, K.D. (1963). *The Art of Simulation*. London: English Universities Press.

How the behaviour of parasites and predators promotes population stability

DAVID ROGERS and STEPHEN HUBBARD

Hope Department of Entomology, University of Oxford

Simple deductive models for the interaction between predators and parasites and their prey or hosts are often unstable in the absence of any intraspecific population regulation. The introduction of a theoretical parasite or predator into a stable host or prey model based, for example, on the logistic equation can introduce either temporary or permanent instability (Maynard-Smith, 1968). In this paper the arguments for parasites will in most cases apply to predators as well. There are, however, two differences that may be important:

(1) Parasites may waste time re-encountering hosts that are already parasitised. This decreases their searching efficiency when exploitation is appreciable.

(2) Parasites have only to parasitise one host individual to leave (at least) one offspring: they are physically smaller than the hosts they attack. Predators often have to eat many prey to produce one offspring because they are frequently larger than their prey. This suggests that a predator's searching capacity will have to be considerably greater than that of a parasite.

The Eltonian assumptions implicit in the notion 'complexity promotes stability' have been criticised in the light of recent mathematical models in which increased complexity almost always leads to increased instability. May (1973) suggests that a more acceptable interpretation recognises the importance of the invariability of the environment in certain regions e.g. the tropics. The constant tropical environment allows the development of what is

Ecological Stability
edited by M.B. Usher and M.H. Williamson.

potentially a more unstable linking than that found in more seasonal habitats. This reverses the Eltonian concept to one that 'stability promotes complexity'.

At the moment there is no question of choosing between these two ideas. The relevant information is not yet available. But it seems appropriate to keep them both in mind when considering how parasite and predator species may have evolved, and particularly when considering the restriction or extension of the number of host or prey species attacked. If, during the course of evolution, the number of host or prey species is halved, how does this affect the stability of the populations involved? Can any features of parasite behaviour be related to the number of host species attacked? If so, can the changes be given an acceptable interpretation?

The present paper takes a simple insect parasite-host model as a paradigm of a much wider range of mathematical models. It then examines the assumptions of the model in the light of recent field and laboratory information. In the present case various features of natural insect parasites differ from those assumed in the model in ways that tend to decrease the instability of the interaction. This could be an example of May's suggestion that although in general the complex models are less stable than the simple ones, the natural situation is not a 'general' case. It is, rather, mathematically atypical. The conclusion is that if we try to model ecosystems solely from the standpoint of mathematics we are searching for the 'needle in the haystack'.

The short cut to the solution is to examine the behaviour of the animals involved that ultimately determines the strength and variability of their interactions with each other. The remainder of the paper suggests how these new features of parasite behaviour are apparently related to each other in an adaptive way.

A SIMPLE PARASITE-HOST MODEL

We take as our example Nicholson's model (Nicholson and Bailey, 1935), the ideas of which owe a lot to Fiske's (1910) work. Nicholson assumed that insect parasites are limited by their searching capacity (an idea originally appreciated by Muir, 1914) and that each individual searche a constant proportion of the host area independently of all other parasites in the area. The number of hosts attacked

(N_{ha}) in an area containing N hosts and P parasites is found from

$$N_{ha} = N\{1 - \exp(-aP)\} \tag{1}$$

where a is a constant reflecting the searching efficiency of the parasite. It was called by Nicholson the parasite's 'area of discovery' and can be visualized as that proportion of the total area that the parasite searches during its lifetime.

Once such a very definite statement has been made, certain conclusions logically follow. We can, for example, investigate the behaviour of the model in predicting or describing population changes over a number of generations (Burnett, 1958a; Varley and Gradwell, 1963). When we do this we find that the models are unstable: changes in parasite numbers lag behind changes in host numbers so that host mortality in any generation is not directly related to host density (as it is in the logistic equation), but is instead a function of parasite density. Fairly obviously such models can be stabilised by the introduction of other mortalities whose effects are directly related to host population density. Perhaps more unexpectedly they can also be stabilised by direct density dependent mortalities acting on the parasite population itself.

Historically such modelling attempts led to the general conclusion that direct density dependent mortalities promote stability whilst delayed density dependent mortalities - caused by specific parasites and predators - promote unstable oscillations in population density (e.g. Varley, 1970). It remained to be shown whether or not natural parasites and predators behaved in the ways the model assumed.

Population oscillations occur in simple models like Nicholson's for two major reasons:

(1) Parasite searching behaviour was assumed to be independent of host and parasite density and distribution.

(2) The parasites were assumed to be specific, i.e. restricted to one host species.

These two assumptions have often been made for the sake of mathematical convenience. They were not based on biological observations.

THEORY VERSUS REALITY: NON-RANDOM PARASITE BEHAVIOUR

From an evolutionary point of view it must be assumed that any behaviour on the part of an enemy species which increases the production of its offspring in the next generation will be selected for. At first sight therefore we might expect predators and parasites to concentrate their attention in regions of high prey or host density: and we would expect them to be polyphagous. When either or both of these happen, there will be an increase in offspring production. Immediately we can see a contradiction between what we would expect a parasite or predator to do and what simple models say that they do. Thus Nicholson's model assumed that parasitic and predatory insects search for their hosts or prey at random. But the observation that they do not is not a new one. In fact we can go back to one of the earliest works of natural history and find evidence that the significance of non-random search by insects was appreciated even centuries ago.

'The Bestiary' is a serious work of natural history which finally achieved its present form in the 12th Century A.D. (White, 1960) although oral and written tradition date back to before the 5th Century B.C. It describes, among various accounts of Griffins and Gorgons, the behaviour of ants that whilst foraging 'go along the tracks (of loaded ants) to the place where they found corn, and they carry back their own grain to the nest. Mere words, you see, are not an indication of being provident. Provident people, like ants, betake themselves to that place where they will get their future reward' (White, 1960, p.97).

The basic idea is certainly there in the original, but lot of the charm is sadly lost in the translation to today' language - that ants aggregate in regions of high food density. Six examples of non-random search by both insect predators and parasites are shown in Fig.1. The expectatio for each graph on a random search basis is that the points will fall on a horizontal straight line. In each case they do not. The predators or parasites have aggregated, and so increased their efficiency of search.

It is possible to try to model this behaviour at a number of different levels (Hassell and Rogers, 1972; Murdi and Hassell, 1973; Hassell and May, in press). Previously Hassell and Rogers (1972) showed that aggregation within each host generation could by itself decrease the instability of a Nicholsonian model, because hosts in low density areas are effectively protected from attack. The aggrega-

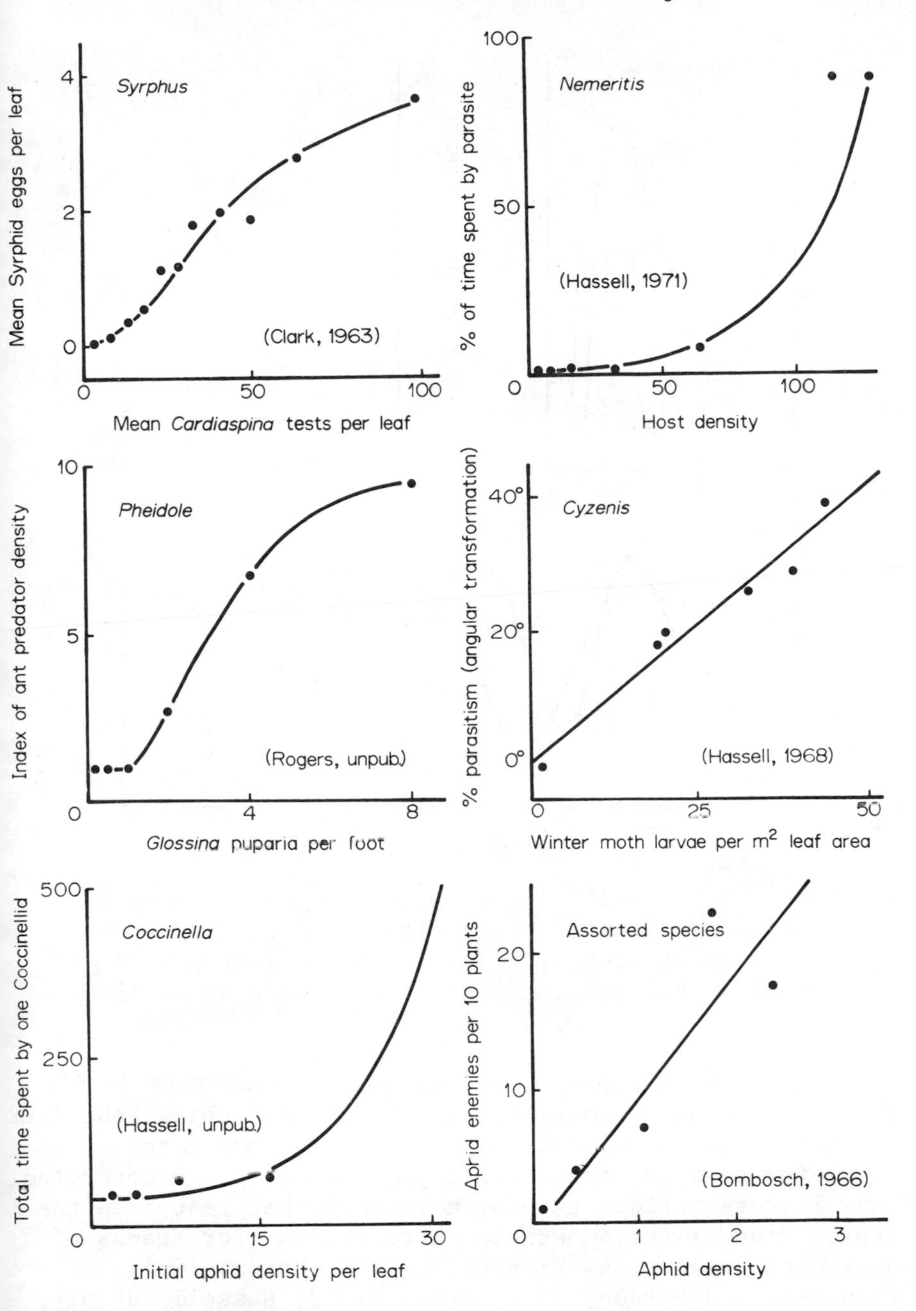

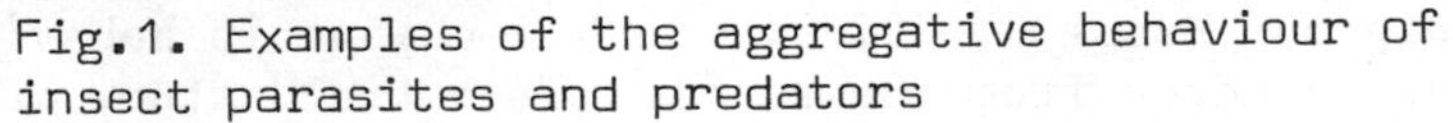
Fig.1. Examples of the aggregative behaviour of insect parasites and predators

ion response used, and the outcome of the model, are ill-
strated in Fig.2.

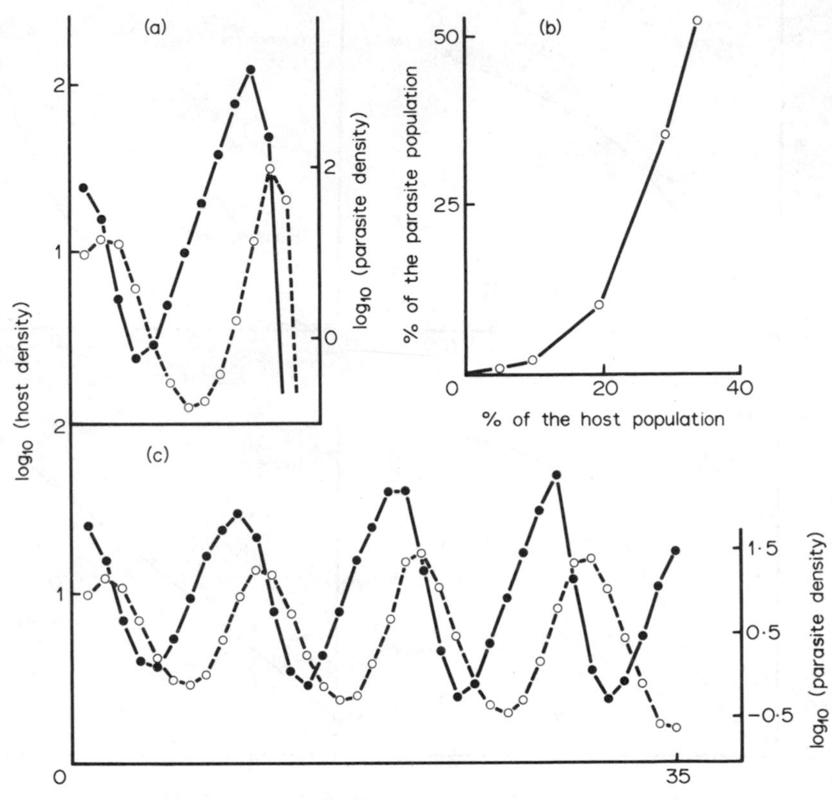

Fig.2. The stabilising effect of aggregation on a simple Nicholsonian model, (a) model with parasites searching at random, (b) aggregative response of the parasite included in the model shown in (c). (From Hassell and Rogers, 1972)

Fig.2 models aggregation only at an elementary level. Ideally we want to go right back to the searching behaviour of individual parasites, to see if we can model this. We know, for example, that once a prey or host is encountered, many insects perform turning movements that result in the exploration of the immediate surroundings (for tracks of searching insects see Fleschner, 1950; Hafez, 1961; Richerson and Borden, 1972; Evans, 1973; Hassell and May, 1973; Murdie and Hassell, 1973). Such a change in behaviour can also be detected from the results of a field study of predation on tsetse fly puparia by *Pheidole* (Fig.1). The puparia, produced from a laboratory colony of adults, were buried in a natural situation along the edges of 25 foot squares and at a depth of one half to one inch. The spacing between the puparia was varied between one and a half and 36 inches in different replicates of the experiment (14 replicates at each density) and the position of each puparium was marked with a short piece of wire. The puparia were dug up after a fortnight, long before the adults would have emerged, and the sequences of healthy or predated puparia were scored for each replicate. Ants of the genus *Pheidole* appeared to be mainly responsible for the observed predation. The experimental results were analysed using the one-sample runs test (Siegel, 1956) which determines whether the puparia were discovered in a purely random sequence, or whether there was some connection between the fate of one puparium and those of its neighbours. Table 1 illustrates the working of this test for a hypothetical example taken from everyday experience, and Fig.3 shows the result of the experiment on the puparia. z, under the null hypothesis, is normally distributed with zero mean and unit standard deviation. At high puparial

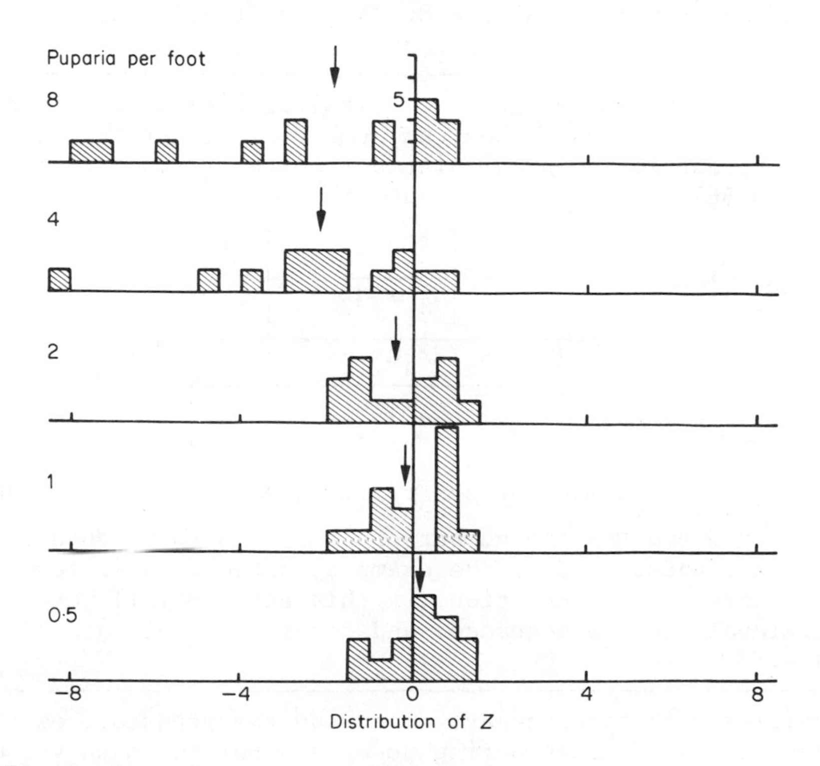

Fig.3. Distribution of z in experiments on the predation of Tsetse fly puparia at different densities. Arrows indicate mean values

Table 1. Examples of the one-sample runs test to illustrate variation in the calculated statistic z. The examples are queues at the cinema

(1) Evening queue at the box office:

M F M F M F M F M F M F M F M F

No. of males = 8, females = 8. No. of 'runs' = 16, $z = +3{\cdot}62$

(2) Queue for icecream during the intermission:

MM FFF MMM F M F M FFF M

No. of males = 8, females = 8. No. of 'runs' = 9, $z = 0{\cdot}0$

(3) School classes queue for afternoon matinee:

FFFFFFFF MMMMMMMM

No. of males = 8, females = 8. No. of 'runs' = 2, $z = -3{\cdot}62$

For small samples the level of significance is usually read from special tables. When the number of either type is greater than 20, z is calculated from the formula (Siegel, 1956, p.56):

$$z = \frac{r - \left[\dfrac{2n_1n_2}{n_1 + n_2} + 1\right]}{\left[\dfrac{2n_1n_2(2n_1n_2 - n_1 - n_2)}{(n_1 + n_2)^2(n_1 + n_2 - 1)}\right]^{0\cdot5}}$$

where n_1 = the number of one type, n_2 = the number of the other type and r = the number of 'runs' in the sequence (runs are underlined in the examples above). z is zero when there is no connection (in this case social) between individuals in the sequence, and non-zero in all other cases.

densities z is often negative because the predators easily found series of neighbouring puparia. But the mean value of z approaches zero as the spacing was increased and the puparia were discovered at random.

THEORY VERSUS REALITY: INTERFERENCE BETWEEN SEARCHING PARASITES

Considerable aggregation has its disadvantages, the main one being that it leads to local over-exploitation of the prey or host population. This might occur to such an extent as to make a previously profitable area no longer worth investigating because the density of prey or healthy hosts has diminished to a low level.

It now appears that interactions between searching individuals reduce this tendency towards over-exploitation by increasing the chance of local emigration. This has been called 'interference' (Hassell and Varley, 1969) and is measured in the laboratory as a reduction in the average searching efficiency of the individual as density increases (Edwards, 1961; Hassell, 1971). The average efficiency is reduced because as predator or parasite density increases the number of non-searching individuals also increases (Hassell, 1971). They usually rest on the walls of their cages, or even try to escape altogether.

Let us regard such an experiment as a sort of equilibrium between searching and non-searching individuals, in order to find out what happens in a short interval of time Δt.

Let P be the total number of parasites present and let P_s be the number searching at the start of Δt. We will assume that an encounter between parasites leads to a period of time T_w being wasted by each. During the time Δt the number of encounters, E, made by each parasite is found from

$$E = b(P_s - 1)\Delta t \tag{2}$$

where b represents the rate of encounter between parasites. Equation 2 states simply that the number of encounters made by one parasite is related to the density of other parasites searching in the same area. The total number of encounters made by all parasites during Δt is therefore

$$E = bP_s(P_s - 1)\Delta t \tag{3}$$

which in this simple model is equal to the number of parasites discontinuing search during Δt, P_l. The number of non-searching parasites that resume search during Δt, P_r, is described by

$$P_r = (P - P_s)\Delta t/T_w \tag{4}$$

Once equilibrium is achieved

$$P_l = P_r \tag{5}$$

and therefore

$$bP_s(P_s - 1) = (P - P_s)/T_w \tag{6}$$

from which

$$T_w bP_s^{\ 2} + (1 - T_w b)P_s - P = 0 \tag{7}$$

a quadratic that can be solved for P_s (Rogers and Hassell, in press). If Q is the searching efficiency of one of the P_s parasites then the average searching efficiency, a, of each of the P parasites is

$$a = QP_s/P \tag{8}$$

Sometimes the following approximation will hold

$$P_s/P \simeq 1/P^m \tag{9}$$

where m is a constant such that $0 < m < 1$. Combining Equations 8 and 9 gives

$$a = QP^{-m} \tag{10}$$

which is the linear model for searching efficiency that is usually applied to experimental results (Hassell, 1971). Although it is not strictly accurate it is mathematically quite convenient, and we will return to it later.

Q, called the 'quest constant' by Hassell and Varley (1969), represents the efficiency of the parasite at unit population density, i.e. when $P_s = P = 1$. By assigning values to $T_w b$ in Equation 7 it is possible to investigate how the average searching efficiency changes with parasite density. When this is done a series of curves is obtained (Fig.4 upper graph), such curves becoming virtually linear at high values of $T_w b$. An inspection of published interference relationships (Hassell, 1971, Hassell and Rogers, 1972) shows the same family of curves as the simple model predicts (Fig.4, lower six graphs).

Interference between parasite individuals increases with

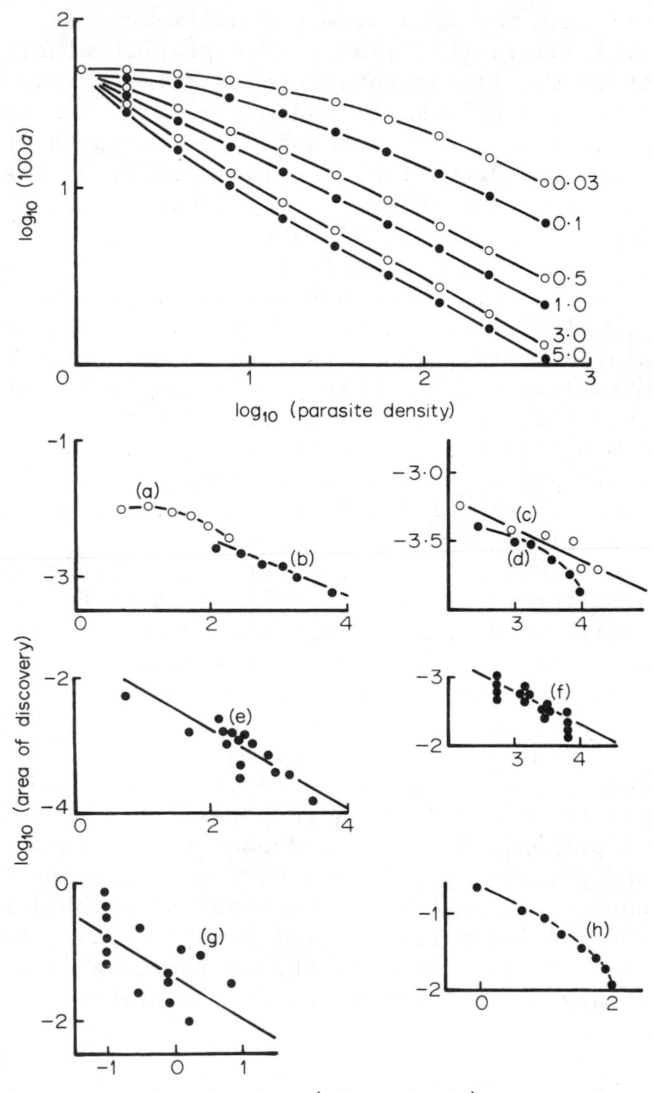

Fig.4. A comparison of the predictions of the simple behavioural model for interference for varying values of $T_w b$ (upper graph), with some published relationships for real parasites (lower graphs), which are (a) *Dahlbominus*, (b) *Encarsia*, (c) *Cryptus*, (d) *Chelonus*, (e) *Nemeritis*, (f) *Pseudeucoila*, (g) *Cyzenis*, and (h) *Trichogramma*. Details are given in Table 2

density and affects the efficiency of the parasite population in the same way as a direct density dependent mortality (Hassell and Varley, 1969). The precise stability properties of the linear interference relationship have been investigated by Hassell and May (1973). As the slope, m, of the interference relationship is increased from zero (the Nicholsonian situation) the instability of the model is initially reduced. However, beyond a certain value for m, that depends on the hosts' rate of reproduction, the parasites are no longer capable of reducing the host population. For example, when $m = 1{\cdot}0$ the efficiency of the parasite population is unaffected by changes in the parasite population size. The parasites cause a constant percentage mortality and the host population increases without limit. Thus there is only a relatively narrow range of values of m within which a stable interaction with the host population is possible (Hassell and May, 1973). That most of the known values for m fall within this range may be more than a coincidence.

The upper limit to parasite searching efficiency implied by the curvilinear relationships in Fig.4 is bound to affect stability adversely. But the number of parasite generations 'spent' in this horizontal part of the curve will be determined by a number of factors such as the value of the innate searching capacity, Q, and the hosts' reproductive rate (Rogers and Hassell, in press).

Interference, which is measured in an enclosed space in the laboratory as a reduction in parasite searching efficiency, will probably in the field encourage a re-distribution of the parasite population through migration to previously unexplored regions of the host distribution. Thus aggregation and interference will be dynamically balanced, although the precise level of this balance is likely to depend on many factors and to be of evolutionary significance.

RELATIONSHIPS BETWEEN THE PARAMETERS OF PARASITE SEARCH

In the simple behavioural model described in the previous section it is possible to vary the rate of encounters between parasites completely independently of the efficiency in the absence of interference. But for natural parasites it is likely that those species with a high searching efficiency, Q, will encounter each other more often than those with a lower efficiency. In other words as Q in-

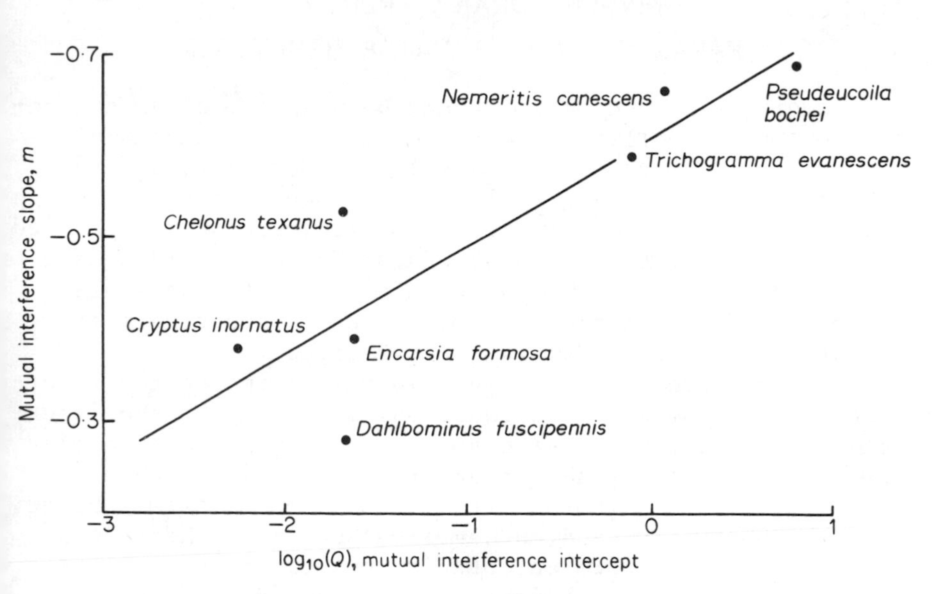

Fig.5. The relationship between the mutual interference constant, m, and the quest constant, Q, for several species of insect parasite. The equation of the line is $Y = 0{\cdot}117X - 0{\cdot}612$

creases the rate of encounters between parasites (b in the model) will also increase. As b increases it is likely that the amount of interference as measured by m (Equations 9 and 10) will be greater. We would therefore expect a relationship between the searching efficiency, Q, and the slope of the interference equation, m. So far we only have data from laboratory experiments for working out this relationship, but the trend is as expected (Fig.5). In Fig.5 the various results have been corrected so that Q is the efficiency in square metres per parasite per day (Rogers and Hassell, in press).

Thus the simple model for interference described here potentially provides an infinite variety of combinations of Q and m for any one species (i.e. Q and $T_w b$ infinitely variable). But the values of these parameters for several species of insect are related to each other in an intelligible way. This is an idea that we shall consider again.

EVOLUTIONARY TRENDS IN PARASITE AND PREDATOR BEHAVIOUR

Both aggregation and interference between parasites can decrease the instability of the simple Nicholsonian model (Hassell and Rogers, 1972; Hassell and May, 1973 and in press). Interference has also been shown to reduce the egg production of the few invertebrate predators so far studied (Kuchlein, 1966; Evans, 1973) and so can be regarded as a general phenomenon. This section explores the possible evolution of interference behaviour as an adaptive feature of the biology of predators and parasites.

We start with the two assumptions that increasing interference encourages population stability and that the interaction of a single parasite species with many host species is inherently more stable than its interaction with a single host species. The first assumption is firmly based on both observation and mathematical models. The second would probably be more acceptable to a biologist than to a mathematician, since it incorporates the idea of spreading the risk. In general we postulate that there should be less variability in the supply of two or more resources

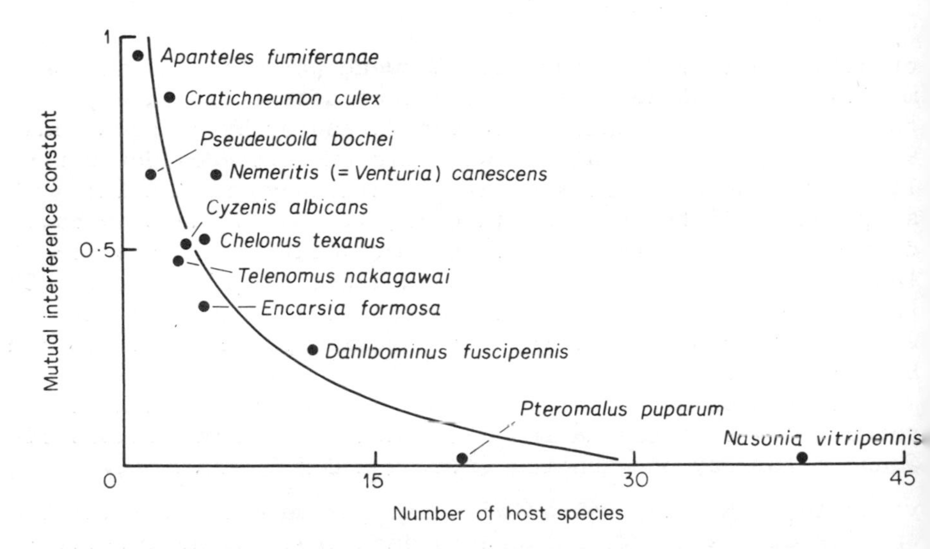

Fig.6. The relationship between the mutual interference constant, *m*, and the number of host species attacked

than in the supply of only one, and that insect parasites primitively operated on this principle.

At the present time different species of insect parasites attach varying numbers of host species. Polyphagous parasites have a great variety of alternative hosts to attack when any single species becomes scarce. But since monophagous species are tied to the abundance of only one host species they may tend to be numerically more unstable. We might therefore expect as a result of selection a greater stabilising force in the behaviour of monophagous species to have a higher interference constant, m, than polyphagous species. This appears to be the case in the few species for which suitable information is available (Fig.6). (The sources of information for Figs. 5 and 6 are given in Table 2).

We do not want to get involved in the theory of the evolution of parasitic and predatory insects and whether primitive forms were monophagous or polyphagous. But we can begin to consider the changes which might occur during the transition in order to answer the question "How might searching efficiency be affected during the move from polyphagy to monophagy?".

The trend towards monophagy (perhaps as a result of competition with other species) will lead to reduced population numbers and possible extinction unless the parasites leave as many (or more) offspring if they became monophagous as they could previously during their polyphagous phase. This requires that searching efficiency increases with the reduction in number of host species attacked. A simple model can describe this. Assume that a polyphagous parasite is in some sort of equilibrium with a total of N_e hosts, which comprise a number of different species of approximately equal abundance. The number of hosts parasitised in each generation, N_{ha}, is found from Equation 1:

$$N_{ha} = N_e(1 - \exp\left[-aP_e\right]) \quad (11)$$

where P_e is the density of parasites and a their area of discovery. If the parasite species is now restricted to a proportion r of the host species (and therefore to rN_e hosts), and if its searching efficiency increases to a', parasite density will remain constant if

$$N_e(1 - \exp\left[-aP_e\right]) = rN_e(1 - \exp\left[-a'P_e\right]) \quad (12)$$

This equation can be used to predict how the new searching efficiency, a', would have to change with r. This is shown

Table 2. Biological parameters for species shown in Figs.5 and 6

Species	Number of host species attacked	log Q m^2 parasite^{-1} day^{-1}	m	Field/ Lab.	Original Source
Apanteles fumiferanae Vier.	1	?	0·96	Field	Miller (1959)
Chelonus texanus Cress.	5	-1·70	0·53	Lab.	Ullyett (1949a)
Cryptus inornatus Pratt.	5	-2·25	0·38	Lab.	Ullyett (1949b)
Cyzenis albicans (Fall.)	4	?	0·52	Field	Hassell and Varley (1969)
Cratichneumon culex (Muell.)	3	?	0·87	Field	G.R. Gradwell (unpub.)
Dahlbominus fuscipennis (Zett.)	12	-1·66	0·28	Lab.	Burnett (1956)
Encarsia formosa Gahan	2	-1·62	0·39	Lab.	Burnett (1958b)
Pseudeucoila bochei Weld.	2	0·80	0·69	Lab.	Bakker *et al.* (1967)
Nasonia vitripennis (Walker)	38	-1·56	0·02	Lab.	Edwards (1961)
Nemeritis canescens (Grav.)	8	0·05	0·66	Lab.	Hassell (1971)
Pteromalus puparum (L.)	20	?	0·00	Lab.	Hamilton (1926)
Telenomus nakagawai Watanabe	3	?	0·48	Field	Nakasuji *et al.* (1966)
Trichogramma evenescens (Oliv.)	?	-0·12	0·59	Lab.	Edwards (1961)

The number of host species were obtained from Thompson, W.R., *A Catalogue of the Parasites and Predators of Insect Pests*, The Commonwealth Institute of Biological Control, in several parts; Muesebeck, C.F.W. *et al.* (1951), *Hymenoptera of America North of Mexico*, U.S.D.A., Agricultural Monograph No.2, with Supplements 1 by Krombein, K.W. (1958) and 2 by Krombein, K.V. and Burks, B.D. (1967). The Memoirs of the American Entomological Institute were consulted for synonyms.

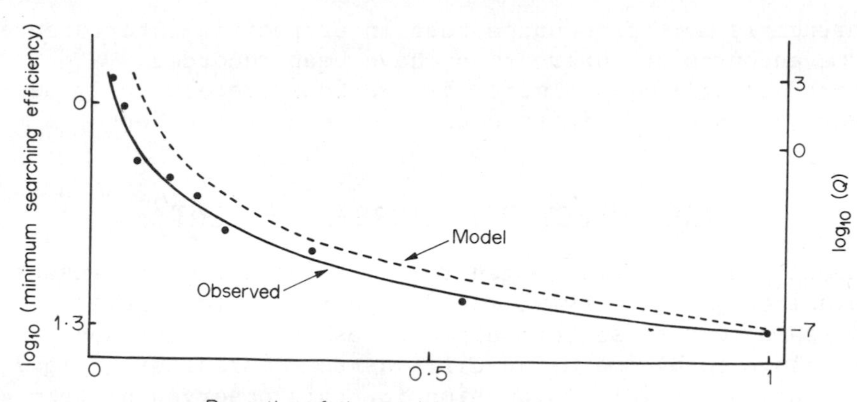

Fig.7. A comparison of the change in searching efficiency with the change in the number of host species attacked, as predicted by the model and as observed from the data in Figs.5 and 6. The upper curve is the model, the lower curve fitted to observed data

in Fig.7 ('Model' curve). Fairly obviously the searching efficiency increases during the trend towards monophagy. But it does not increase linearly. The figure shows that a parasite that becomes restricted to a single host species must have a disproportionately greater search efficiency than a polyphagous parasite species. It is now possible to compare this figure with a combination of the two previous figures which apply to real parasites. Figs.5 and 6 together provide the 'Observed' curve of Fig.7, which is similar to that of the Model. This result again illustrates that models could provide an infinite variety of combinations of searching parameters and host range. Yet nature has not selected these at random.

One important point comes from a comparison of Figs.6 and 7 which relate m and Q respectively to the number of host species attacked. It is obviously possible that the higher searching efficiency of monophagous species automatically leads to increased interference. But it is also likely that the effect of an encounter between searching parasites, in terms of time wasted, has experienced a certain amount of selection. Thus if no time is wasted after an encounter, then there will be no interference (m = 0), but the interaction with the host species would be unstable. Could there have been selection of the extent of interference: just enough to stabilise the interaction? Support for this idea would come from observations that intraspecific interference is more pronounced than interspecific interference. Some encouraging observations have been recorded (Spradbery, 1970) but more observations are obviously necessary before a definite conclusion can be reached.

ARE INSECT PREDATORS DIFFERENT?

Insect predators are in general less specific than insect parasites. In theory we would expect them to experience the same sort of selection pressures as parasites and therefore to become as specific as some parasites. Fig.7 suggests a partial explanation for this observed difference. In the introduction we suggested that predators will have higher searching capacitites than parasites because they consume a large number of prey during their developmental and reproductive periods. In other words they will initially start at a higher point in Fig.7 than that shown for the theoretical parasite. A trend towards monophagy from this primitive condition might require a physically impossible increase in their searching capacity, the increase being disproportionately related to the decrease in the number of species eaten. Only by attacking a large number of species might predators obtain enough food to survive.

CONCLUSION

The evolutionary move from polyphagy to monophagy, or the reverse, is apparently associated with adaptive changes in the parameters of parasite and predator search. They are adaptive in the sense that they increase the chances of survival both in the short term (e.g. a high coefficient of search) and in the long term (a mutual interference constant that ensures stability). Within any particular parasite or predator generation the efficiency of the individual is increased by its tendency to spend more time in regions of high host or prey density. There will be a dynamic balance between aggregation of the searching population and the increased interference that is bound to occur. Selection may determine the quantitative interaction between the two.

Finally these examples illustrate that even simple models can provide a kaleidoscope of metaphors for the natural situation. Only observations of living parasites and predators will decide which is correct.

ACKNOWLEDGEMENTS

We would like to acknowledge the helpful and stimulating discussions we have had with Professor G.C. Varley of the Hope Department and Dr. Mike Hassell of Imperial College. They have both made us think more deeply about the implications of our observations.

SUMMARY

(1) A simple parasite-host model is examined in the light of recent observations and experiments on insect parasites and predators.

(2) Whereas the original deductive model predicts an unstable interaction between the two populations, the incorporation of observed features of the behaviour of real insect parasites decreases the instability appreciably.

(3) Whilst a multiparameter model for parasite-host interaction has an almost infinite variety of combinations of parameter values, the natural situation appears to be less complicated and more elegant. Adaptive components of parasite behaviour are related to each other in an intelligible way.

(4) The Conclusion suggests that useful progress in mathematical modelling will occur only if reference is continually made to the natural situation.

REFERENCES

Bakker, K., Bagchee, S.N., Van Zwet, W.R. and Meelis, E. (1967). Host discrimination in *Pseudeucoila bochei* (Hymenoptera: Cynipidae). *Entomologia exp. appl.*, 10, 295-311.

Bombosch, S. (1966). Distribution of enemies in different habitats during the plant growing season. *Ecology of Aphidophagous Insects* (Ed. by I. Hodek), pp.171-5 The Hague: Junk Publishers.

Burnett, T. (1956). Effects of natural temperatures on oviposition of various numbers of an insect parasite (Hymenoptera, Chalcididae, Tenthredinidae). *Ann. ent. Soc. Am.*, 49, 55-9.

Burnett, T. (1958a). A model of host-parasite interaction. *Proc. 10th. int. Cong. Ent. (1956)*, 2, 679-86.

Burnett, T. (1958b). Dispersal of an insect parasite over a small plot. *Can. Ent.*, 90, 279-83.

Clark, L.R. (1963). The influence of predation by *Syrphus* sp. on the numbers of *Cardiaspina albitextura* (Psyllidae). *Aust. J. Zool.*, 11, 470-87.
Edwards, R.L. (1961). The area of discovery of two insect parasites, *Nasonia vitripennis* (Walker) and *Trichogramma evanescens* Westwood, in an artifical environment. *Can. Ent.*, 93, 475-81.
Evans, H.F. (1973). A study of the predatory habits of *Anthocoris* species (Hemiptera - Heteroptera). *Unpublished D. Phil. thesis, Oxford.*
Fiske, W.F. (1910). Superparasitism: an important factor in the natural control of insects. *J. econ. Ent.*, 3, 88-97.
Fleschner, C.A. (1950). Studies on searching capacity of the larvae of three predators of the citrus red mite. *Hilgardia*, 20, 233-65.
Hafez, M. (1961). Seasonal fluctuations of population density of the cabbage aphid *Brevicoryne brassicae* (L.) in the Netherlands, and the role of its parasite, *Aphidius (Diaretiella) rapae* (Curtis). *Tijdschr. PlZiekt.*, 67, 445-548.
Hamilton, A.G. (1926). Biology of *Apanteles glomeratum* and *Pteromalus puparum* parasitising *Pieris brassicae*. *Diploma thesis, Imperial College.*
Hassell, M.P. (1968). The behavioural response of a tachinid fly (*Cyzenis albicans* (Fall.)) to its host, the winter moth (*Operophtera brumata* (L.)). *J. Anim. Ecol.*, 37, 627-39.
Hassell, M.P. (1971). Mutual interference between searching insect parasites. *J. Anim. Ecol.*, 40, 473-86.
Hassell, M.P. and May, R.M. (1973). Stability in insect host-parasite models. *J. Anim. Ecol.*, 42, 693-726.
Hassell, M.P. and May, R.M. (in press). Aggregation of predators and insect parasites and its effect on stability. *J. Anim. Ecol.*, 43.
Hassell, M.P. and Rogers, D.J. (1972). Insect parasite responses in the development of population models. *J. Anim. Ecol.*, 41, 661-76.
Hassell, M.P. and Varley, G.C. (1969). New inductive population model for insect parasites and its bearing on biological control. *Nature, Lond.*, 223, 1133-7.
Kuchlein, J.H. (1966). Mutual interference among the predacious mite *Typhlodromus longipilus* Nesbitt (Acari, Phytoseiidae). I. Effects of predator density on oviposition rate and migration tendency. *Meded. Rijksfac. LandbWet. Gent.*, 31, 740-6.
May, R.M. (1973). *Stability and Complexity in Model Eco-*

systems. Princeton: Princeton University Press.
Maynard Smith, J. (1968). *Mathematical Ideas in Biology*. London: Cambridge University Press.
Miller, C.A. (1959). The interaction of the spruce budworm, *Choristoneura fumiferana* (Clem.), and the parasite *Apanteles fumiferanae* Vier. *Can. Ent.*, 91, 457-77.
Muir, F. (1914). Presidential address. *Proc. Hawaiian Ent. Soc.*, 3, 28-42.
Murdie, G. & Hassell, M.P. (1973). Food distribution, searching success and predator-prey models. *The Mathematical Theory of the Dynamics of Biological Populations*. (Ed. by M.S. Bartlett and R.W. Hiorns), pp.87-101. London: Academic Press.
Nakasuji, F., Hokyo, N. and Kiritani, K. (1966). Assessment of the potential efficiency of parasitism in two competitive scelionid parasites of *Nezara viridula* L. (Hemiptera: Pentatomidae). *Appl. Ent. Zool.*, 1, 113-9.
Nicholson, A.J. and Bailey, V.A. (1935). The balance of animal populations. Part I. *Proc. zool. Soc. Lond.*, 551-98.
Richerson, J.V. and Borden, J.H. (1972). Host finding of *Coeloides brunneri* (Hymenoptera: Braconidae). *Can. Ent.*, 104, 1235-50.
Rogers, D.J. and Hassell, M.P. (in press). General models for insect parasite and predator searching behaviour: Interference. *J. Anim. Ecol.*, 43.
Siegel, S. (1956). *Nonparametric Statistics*. New York: McGraw-Hill.
Spradbery, J.P. (1970). Host finding by *Rhyssa persuasoria* (L.), an ichneumonid parasite of siricid woodwasps. *Anim. Behav.*, 18, 103-14.
Ullyett, G.C. (1949a). Distribution of progeny by *Chelonus texanus* Cress. (Hymenoptera: Braconidae). *Can. Ent.*, 81, 25-44.
Ullyett, G.C. (1949b). Distribution of progeny by *Cryptus inornatus* Pratt (Hymenoptera: Ichneumonidae). *Can. Ent.*, 81, 285-99.
Varley, G.C. (1970). The need for life tables for parasites and predators. *Concepts of Pest Management* (Ed. by R.L. Rabb and F.E. Guthrie), pp.59-70. Raleigh: North Carolina State University.
Varley, G.C. and Gradwell, G.R. (1963). The interpretation of insect population changes. *Proc. Ceylon Assn. Adv. Sci.*, 18, 142-56.
White, T.H. (1960). The Bestiary - a Book of Beasts. New York: Capricorn Books, G.P. Putnam's Sons.

Part three: Temporal studies between trophic levels: predator–prey systems

The kinetics of polyphagy

JOHN R.W. HARRIS
Department of Biology, University of York

INTRODUCTION

The dynamics of an ecosystem will be influenced by the number of species which it incorporates, but also by the form of trophic web which links them (MacArthur, 1955; Watt, 1965). The extent to which each species feeds upon each other, if at all, will be important in determining the character of the system. Fundamental to elucidating these links and the way in which they may vary is the understanding of the functional response (Solomon, 1949) of individual predators to changes in the density of their various prey. The purpose of this paper is to present one approach to this problem.

From the outset we will define the two variables which it is wished to relate. The density of prey is the mean number of individuals per unit of the space from which the predator obtains its prey. The rate of predation is taken to be the number of prey consumed per unit time per predator, measured over a period during which no pertinent variable external to the predator, including prey density, changes. The so-called exploitation component of predation (see for example Holling, 1961) is in this way specifically excluded from effect upon it.

Two relations have been widely used between predation rate and prey density. These are illustrated by the two equations given by Rashevsky (1959, Equations 2 and 26)

$$r = R(1 - e^{\xi p}) \quad (1)$$

Ecological Stability
edited by M.B. Usher and M.H. Williamson.

$$r = Rp/(1 + \xi p) \quad (2)$$

in which p is the prey density, r the rate of food consumption (in units of mass per unit time, R the maximum rate and ξ a constant. Relations of the form of Equation 1 have been used by Gause (1934), Ivlev (1961) and, in a more sophisticated model, by Watt (1959). Equation 2 is equivalent to the 'disc' equation of Holling (1959), of which a stochastic version has been proposed by Mertz and Davies (1968) for cannibalism in *Tribolium* species. Rashevsky (1959) argued that Equation 1 would apply to predation which may be regarded as a progressive filling of the digestive tract. He contrasted this non-stationary state with the stationary state in which elimination balances uptake to which he considered Equation 2 to apply.

These relations treat a situation with apparently homogeneous prey. Sukhanov and Shapiro (1971) have extended Equation 1 to the situation of a predator exposed to a number of types of prey. The increased specialisation with increased prey abundance, found by Ivlev (1961) for carp was incorporated by postulating the probability of capture of a given prey once encountered to increase or decrease linearly with increased gut content. This may provide a true representation of a predator which feeds by periodic (for example daily) cropping of small prey over fixed periods of time, interspersed with periods without food, but alternatives are required when feeding rates are more stable or the predator is satiated by individual items of prey.

Holling has extended his basic equation to incorporate the effects of hunger (Holling, 1966) and predator learning when exposed to two types of prey (Holling, 1965). In what follows we will suggest a means of extending an equation of the form of Equation 2 to describe a situation in which a simple predator is exposed to any number of prey types. The development is presented in four distinct stages. First, to introduce the form of the conceptual model to be employed, feeding by an individual predator (or predators at a fixed density) upon prey of a single homogeneous type is considered, an equation equivalent to Equation 2 being derived. The model is next extended to cover the situation of a predator exposed to any number of distinct types of prey at the same time, assuming however that 'satiation' of the predator is always in respect of all prey. This assumption is considered unrealistic and subsequently dropped. The third step, extending the model to allow this introduces considerable generality so that finally restrictions are introduced to produce a form simple enough to relate to observed data.

To maintain flexibility in the model an attempt has been made to retain a simple conceptual image of the system throughout, and to use parameters amenable to distinct biological interpretation. The final equations are couched in terms which might be readily measured by simple experiments, although it must be remembered that expressions involving rates of feeding will require integration if prey densities vary during the period of their estimation (this has been discussed for a situation with one prey by Rogers, 1972).

The 'types' of prey referred to are intended to indicate any groupings of prey which may be usefully distinguished, be they species, size classes, sexes, developmental stages or whatever.

ONE PREDATOR AND ONE PREY TYPE

The prey are conceived as being distributed at random over the space from which they are obtained by the predator. The relative movements of prey and predator during the latter's search, or wait, for food are thought of as random. These premises are incorporated in the assumption that the rate of capture of prey by a hunting predator is proportional to prey density, although this may hold in more situations than are implied by the premises from which it is derived. At a given density of prey it is assumed that the rate of prey capture by hunting predators per unit area is proportional to the density of those predators.

The predatory situation, in which the rate of predation is considered by Holling (1959) limited on the one hand by the rate at which the hunting predator can obtain prey and on the other by the rate at which captured prey is dealt with and the predator returns to the search for further prey, is analogous to the action of an enzyme upon a single substrate. Added to this, Equation 2 is precisely equivalent to the Michaelis-Menten equation relating the rate of an enzyme reaction to the concentration of the enzyme's substrate. This suggests a simple and flexible approach to the development of models of the functional response of predators to the density of their prey.

In a situation in which a predator searches at random for prey randomly distributed in a two-dimensional arena (the selection of an area rather than a volume for discussion is arbitrary and will not affect the results), let the prey be represented by F, with a density of f individuals per unit area, the predator by C. By analogy with a chemical equation we may represent the predation as

$$C + F \xrightarrow{\alpha} CF \xrightarrow{\beta} C$$

where C represents a hunting predator and CF that feeding on or digesting prey. These latter, non-hunting predators, will be referred to as 'satiated' with respect to this prey. α and β are rate constants (or alternatively transition probabilities per unit time) for the changes implied by the arrows. The steady state rate of predation in this system may be derived in a similar manner to that classicaly employed in chemical kinetics, by considering the equilibrium of the numbers of satiated predators. If r is the rate of predation per unit time per predator and R its maximum attainable value, this yields

$$r = \frac{\alpha\beta f}{\alpha f + \beta} \text{ or } \frac{\alpha R f}{\alpha f + R} \qquad (3)$$

since $R = \beta$. Equation 3 is equivalent to Equation 2.

ONE PREDATOR AND ANY NUMBER OF PREY TYPES

This approach can be applied to a situation in which the predator is simultaneously exposed to n types of prey. Using notation as before, the density of the prey F_y is f_y per unit area. In addition, if r_y is the rate of predation by C per unit time per predator on F_y at density f_y in the absence of all other prey and R_y is the maximum attainable rate of predation under these circumstances, we will define u_y as the proportion by which it is possible to increase r_y by an infinite increase in prey density

$$u_y = (R_y - r_y)/r_y$$

Prey equally palatable

If, when the predator is exposed to all n prey types, satiation is assumed always to be in respect of all of them the predation may be represented

$$C + F_y \xrightarrow{\alpha_y} CF \xrightarrow{\beta_y} C$$

as before. In the steady state situation, by again considering the equilibrium of the numbers of predators satiated by each prey type, it may be shown that

$$1/p_y = 1/r_y + (1/r_y - 1/R_y) \sum_{\substack{i=1 \\ i \neq y}}^{n} 1/u_i \tag{4a}$$

where p_y is the rate of consumption of F_y by C per predator per unit time and the r_y and R_y are related to the α_y and β_y by Equation 3. In matrix form

$$\mathrm{t} = \mathbf{A}.\mathrm{b} \tag{4b}$$

where $\mathrm{t} = \{1/p_i\}$, $\mathrm{b} = \{1/u_i\}$ and $\mathbf{A} = \{a_{ij}\}$, for which $a_{ij} = u_i/R_i$ when $i \neq j$, and $a_{ii} = u_i/r_i$. The general form of this relation is indicated in Fig.1, which shows the results obtained from the model in two situations involving two prey types at equal densities. Variations in relative density would be equivalent to varying the relative values of the α_i.

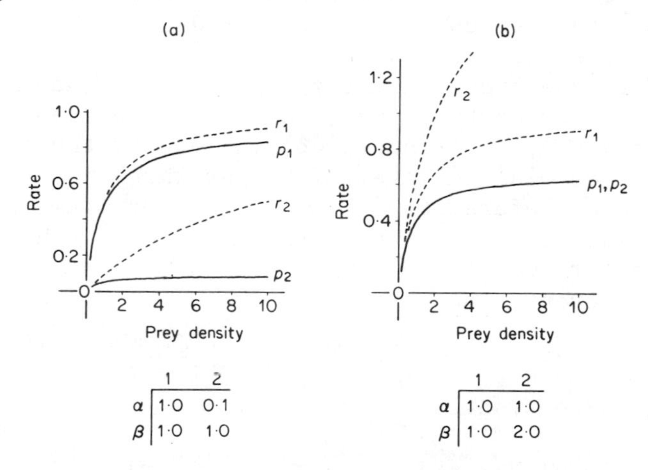

	1	2
α	1·0	0·1
β	1·0	1·0

	1	2
α	1·0	1·0
β	1·0	2·0

Fig.1. The relation of predation rate/predator upon each of two prey types to their densities to be expected under Equation 4, for each alone (r_1, r_2) and in the presence of the other at equal density (p_1, p_2). The respective values of α and β used are as tabulated

To measure the relative rates at which different species of prey are consumed when together in the presence of a predator Ivlev (1961) defined 'electivity', which, in the present notation, is given by

$$e_y = (p_y / \sum_{i=1}^{n} p_i - f_y / \sum_{i=1}^{n} f_i) / (p_y / \sum_{i=1}^{n} p_i + f_y / \sum_{i=1}^{n} f_i)$$

where e_y represents the electivity of the predator for prey F_y. Ivlev found, for fishes feeding on various species of food, electivities tended to diverge as prey densities were increased concurrently. This result is at variance with Equation 4, which gives electivities dependent only on the relative and not the absolute densities of the prey.

Prey of differing palatability

Equation 4 subsumes that the maximum rate of predation is limited by the rate at which the captured prey is dealt with and the predator returns to a state in which it will catch further prey. The time taken to return to this state is assumed to depend only on the prey just captured - not on the prey which is next caught. Such a model would be expected to account for situations in which the rate of predation depended upon the size (or digestability) and ease of capture of the prey. However it is inadequate for cases where the predator shows a preference (dependent upon taste etc.) for certain prey, which need not depend upon ease of capture. Holling (1966) introduced the idea of a 'hunger threshold', a level of hunger which must be reached by a predator before it will attack a prey organism. It would be expected that a predator would have a higher threshold for some (less delectable) prey than for others. Instead of events such as

$$C + F_y \longrightarrow CF_y \longrightarrow C$$

we ought to consider

$$C_j + F_y \xrightarrow{\alpha_{yj}} C_j F_y \xrightarrow{\gamma_{yj}} C_g \xrightarrow{\phi_g} C_{g+1} \xrightarrow{\phi_{g+1}} \dots \xrightarrow{\phi_{j-1}} C_j \xrightarrow{\phi_j} \dots \xrightarrow{\phi_{n-1}} C_n$$

where α_{yj}, γ_{yj} and the ϕ_i again represent rate constants for the events indicated. The prey $F_1, F_2, \dots, F_n$ are arranged in order of decreasing palatability and C_j represents a predator 'hungry' enough to consume any prey

F_y for $y \leq j$. Thus at any time all the predators in a given situation will be considered either completely satiated or else in one of the states C_1, C_2, ... , C_n to which we will refer as 'levels of hunger'. In the steady state the proportion of predators in each condition will be constant. The value of g in the above situation may depend both upon the type of prey which has just been consumed, F_y, and upon the level of hunger prior to this consumption. Clearly, if we imagine hunger to increase along a continuous scale, to which this stepwise model is merely an operational approximation, the predator would not, having consumed a given prey, be expected to return to precisely the level of hunger expressed by the hunger threshold for one of the array of prey types to which it is exposed. To this extent the representation of the rate of conversion of C_g to C_{g+1} by ϕ_g in the above situation must be considered an approximation.

After consuming an individual prey a particular predator may reach any hunger level before capturing further prey, but it will be convenient to divide the predators into groups such that within each the condition at which prey are captured is constant. Although in practice the actual composition of these groups will alter with time the number in each may be treated as constant in the steady state. Within the jth such group we may again consider predation upon F_y to be represented by

$$C_j + F_y \xrightarrow{\alpha_{yj}} C_jF_y \xrightarrow{\beta_{yj}} C_j$$

but now

$$1/\beta_{yj} = 1/\gamma_{yj} + \sum_{i=g}^{j} 1/\phi_i$$

If the predation rate per predator by the jth group upon F_y is denoted by w_{yj} these will be obtained as the p_y from Equations 4. In these the r_{ij} are obtained by substitution of α_{ij} and β_{ij} for α and β in Equation 3. The R_y are replaced by β_{yj} and the u_y by $(\beta_{yj} - r_{yj})/r_{yj}$. From the definition of levels of hunger $\alpha_{yj} = 0$, and hence $w_{yj} = 0$, for all F_y with $y > j$.

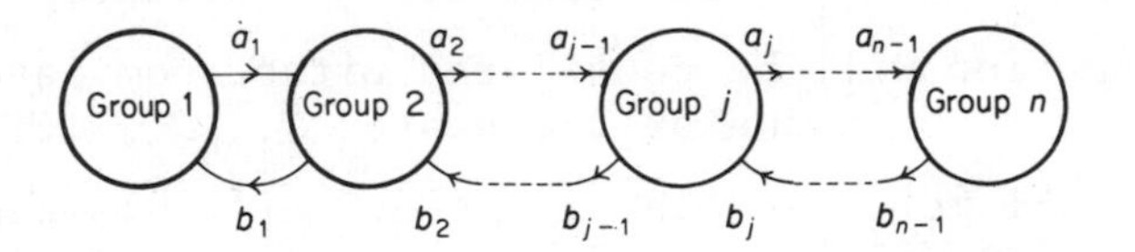

Fig.2. The scheme of 'flows' of predators between notional groups used in the model (see text)

While in particular circumstances alternative schemes may be preferred, in order to obtain the steady state distribution of predators among the notional groups we will here consider those situations which might be represented by Fig.2. a_j represents the rate of transfer per predator per unit time from group j to group $j+1$, and b_j the rate constant for the corresponding transfer in the reverse direction. If the proportion of predators in the jth group be h_j then, in the steady state

$$h_{j+1} = h_j(a_j/b_j) \quad , \quad j > 0 \tag{5}$$

and, since

$$\sum_{j=1}^{n} h_j = 1$$

then

$$h_1 = \left[1 + \sum_{j=1}^{n-1} \prod_{i=1}^{j} a_i/b_i\right]^{-1} \tag{6}$$

It remains to evaluate the a_j and b_j. Taking a predator satiated within the context of the jth notional group to be one which has not yet reached the threshold of hunger for F_j it may be shown that the proportion of unsatiated predators in the group is

$$1 - \sum_{i=1}^{n} w_{ij}/\beta_{ij}$$

The rate of transfer of these, unsatiated, individuals to the $(j+1)$th group is ϕ_j per individual per unit time. Thus the rate of transfer per individual of the jth group, a_j, is given by

$$a_j = \phi_j(1 - \sum_{i=1}^{n} w_{ij}/\beta_{ij}) \tag{7}$$

The value taken by b_j does not follow in a similar deductive manner from our assumptions up to this point. The transfer of which b_j gives the rate represents a decrease in predator hunger. We would expect it to increase as the rate of predation in the $(j+1)$th group increased. A simple possibility is that b_j is formed from a linear combination of the $w_{i,j+1}$. Let

$$b_j = K_j(\sum_{i=1}^{n} k_i w_{i,j+1}) \tag{8}$$

in which the k_i are constant for each F_i and might relate to its 'nutritional value' to the predator. On the other hand K_j is characteristic of the given notional group. We might expect the K_j to be related to the ϕ_j, each relating to the same change in level of hunger. At the simplest, K_j might be proportional to ϕ_j. Taking this to be so the two differ only by a constant multiplier, which may be incorporated in the k_i without loss. So we may postulate $K_j = \phi_j$, giving

$$b_j = \phi_j \; (\sum_{i=1}^{n} k_i \, w_{i,j+1}) \tag{9}$$

Combining Equations 7 and 8, and assuming Equation 9 to apply

$$a_j/b_j = (1 - \sum_{i=1}^{n} w_{ij}/\beta_{ij})/(\sum_{i=1}^{n} k_i w_{i,j+1}) \tag{10}$$

Substitution of this in Equation 6 will yield h_1. Repeated application of Equation 5 for $j = 1, 2, \ldots, n-1$ will give the proportions in each remaining notional group. Then, if

p = $\{p_i\}$, the vector of predation rates on prey F_i,

W = $\{w_{ij}\}$, the square matrix of predation rates by the notional predator group j on F_j, remembering that $w_{ij} = 0$

for $i > j$,

$h = \{h_j\}$, the vector of proportions of the total predator complement comprising the notional groups, we have

$$p = W.h \qquad (11)$$

A restricted model

This provides a rather flexible model within the present framework. It will be assumed that the α_{yj} do not vary with hunger while $y \leq j$. A comparable restriction on the variation in the β_{yj} is obtained by considering a situation in which any prey, once captured, is completely consumed, or at least consumed to a constant extent. In this case we might suppose the time taken for a predator to return to its original state of hunger to vary only with the type of prey and not according to the original state of the predator - assuming a rate of absorption independent of hunger. That is $1/\beta_{yj}$ would be constant over all j for each y. All the β_{yj} may now be designated β_y. The β_y found in the presence of other prey types would not differ from that found with F_y alone, so that, using the notation as before, we can again say $\beta_y = R_y$. Restricting the model in this manner the w_{yj} may be obtained directly as the p_y in Equation 4.

Now the rates of predation per predator by any notional group may be derived from those pertaining when the predator is exposed to each prey alone. However the overall rates of predation per predator in the presence of all the prey will only be completely determined from the additional knowledge of the predation rates for at least $n-1$ different pair-wise combinations of prey types. If Equation 9 applies these will be required to allow estimation of the k_i.

PREFERENCE

Preference is commonly invoked in the discussion of diet. In this context, however, it has been used in a variety of ways. Murdoch (1969) uses the term merely to indicate that the food taken by a predator forms a biased sample of the prey to which it is exposed. Similarly the term introduced by Haussmann (1971) to represent preference in a two-prey extension of an equation of the form of Equation 3 appears effectively to introduce differences in the ease of capture of the two types. In an attempt to eliminate the varying ease of capture of prey Rapport and Turner (1970) have proposed a definition of 'relative preference'. However, within the context of the present model this might be confounded by varying rates of consumption and digestion. Williamson (1957) used the term 'preference factor' to describe a monotonic increasing function of the numbers of one species introducing asymmetry in the action of a common controlling factor on the death rate of a pair of species. Bearing in mind that when the rate of predation depends only upon the ability of the predator to capture and deal with individual items (Equation 4 applies) electivities are constant, independent of overall prey density, it may be seen that this use of preference effectively eliminates these factors from inclusion.

To define the meaning to be attributed to the word here let θ_{ij} be the rate constant corresponding to ϕ_i when the predator is exposed to a combination of F_i and F_j only. In the present context we will restrict phrases such as 'F_i is preferred to F_j' to imply only that $1/\theta_{ij}$ is greater than zero, and the statement that no preference is shown between F_i and F_j to imply only that $1/\theta_{ij}$ equals zero. It will become clear that this use of preference is in fact a particular case of that of Williamson (1957).

PROPERTIES OF THE MODEL SITUATION

Although the situation with only two types of prey will be presented, the general tenor of the implications is unaltered by increasing the number.

Initially we will consider the situation in which the k_i are fixed in value and equal for the two prey, taking them to be unity. A general view of the effect of variations in the density of the two prey is presented by contour plots of the p_1 and p_2 surfaces. In Fig.3 these are plotted for a situation in which the two prey types are identical with respect to α and β, and the asymmetry introduced by preference for one prey (f_1) over the other is clear.

This feature is emphasised by consideration of the

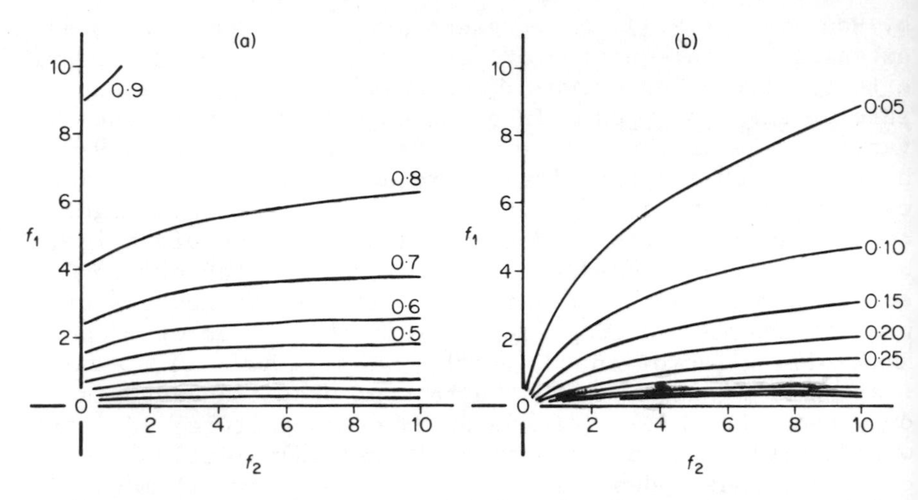

Fig.3. Contour plots in the f_1, f_2 plane of the p_1 (a) and p_2 (b) surfaces, plotted at 0•1 and 0•05 unit intervals respectively. For lack of space p_2 contours above 0•45 are omitted

effect of concomittant variations in prey density, most conveniently when the two densities are equal, the situation already presented in the absence of preference (Fig.1). When two prey types are identical with respect to α and β the rates of predation to which they are subject will be identical while no preference is shown. If this is not the case and one is preferred this is reflected in a divergence of the electivities for the two types as their densities are simultaneously increased (Fig.4a), culminating in a complete specialisation by the predator on the preferred prey. The increase in prey availability reduces the likelihood that a predator will reach the threshold of hunger for the less preferred prey prior to capturing another. At the same time the tendency to reduce the hunger of hungry predators is increased. A similar effect is found when the more readily caught of a pair of prey types is preferred.

In both of these situations we obtain results in qualitative agreement with those found by Ivlev (1961) for carp. MacArthur and Pianka (1966) and Emlen (1966, 1967) each suggest increasing selectivity with increased prey abundance as an energetically optimum strategy for predators. While Emlen implicitly assumes no variation in ease of capture, MacArthur and Pianka specifically assume preference

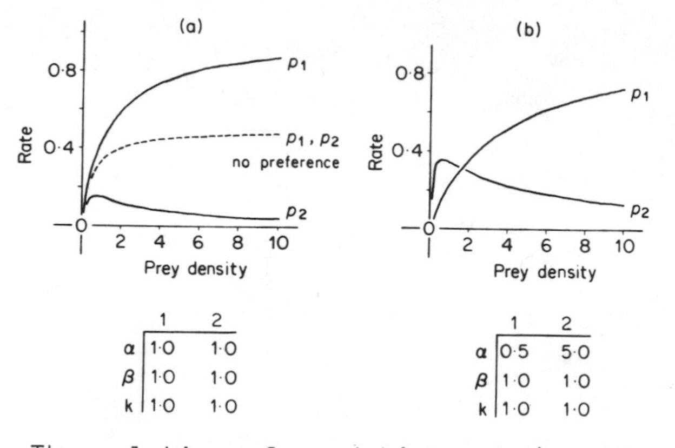

Fig.4. The relation of predation rate/predator upon each of two prey types (F_1 and F_2) to their densities, each in the presence of the other at equal density, to be expected under the restricted version of Equation 11, when F_1 is preferred to F_2. The values of α, β and k are as tabulated

to depend upon ease of capture. In these situations their expected strategy would be approximately achieved by a predator operating according to the present model. Schoener (1969) has considered optimum prey size distributions and it appears from his model that as prey becomes less dense (pursuit or search time increases) the modal prey size should decrease. This would occur under the present model if larger prey were preferred by the predator over smaller individuals.

We may see that, in the simple two-prey situation, such as those described above, the predator will exhibit switching, in the sense of Murdoch (1969), in favour of the preferred type as prey densities increase (Fig.5). This is still true when the value of α for the preferred type is lower than that for the alternative. Now the predator's diet, while showing a higher proportion of the less preferred type at lower densities, will show a preponderance of the other when both are abundant (Fig.4b). Such a situation has been found (personal observation) for the soil mite *Pergamasus longicornis* feeding upon soil collembola. The apparently distasteful *Hypogastrura denticulata* is more easily captured by the mite than more palatable alternatives (*Folsomia fimetaria*, *Sinella caeca*) and the proportion which these latter constitute of its diet shows a marked increase with increased density of equinumerous two-prey populations. The arguments put forward by MacArthur and Pianka (1966) and

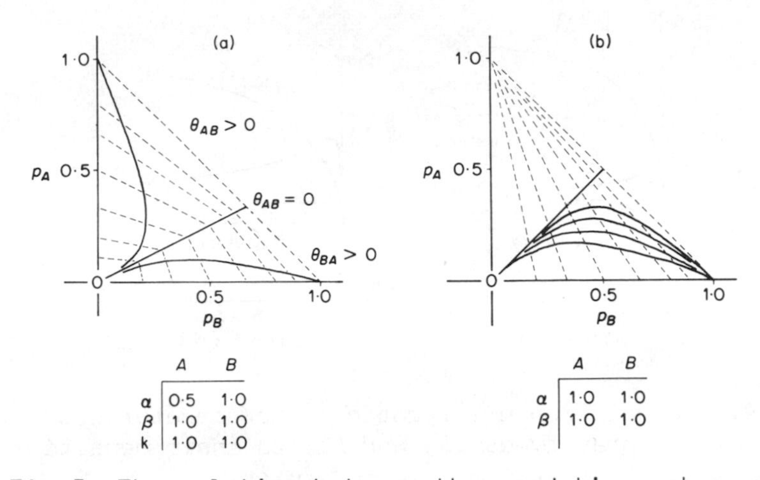

Fig.5. The relation between the predation rates upon two prey types at equal densities showing (a) the effect of preference and (b) the effect of varying k when F_B is preferred to F_A. In (b) the k_i take the values of 0•1 (highest curve), 0•2, 0•4 and 0•8. The dotted lines join points of equal prey density for $f_A = f_B = 0{\cdot}25, 0{\cdot}5, \dots, 8$ and ∞. In (b) they are extended above the 'no preference' line to intersect the p_A axis at β_A to indicate their mode of construction, the P_B axis representing total specialisation in F_B

Emlen (1966, 1967) to specify an optimum diet are based upon the assumption that selection acts upon the predator to maximise its net energy input per unit of hunting time. They do not consider the effect of predator selectiveness upon its prey, nor that the viability of the predator will depend upon the stability (in the sense of both longevity and low variability) of the community of which it is part. For a predator upon a pair of prey responding to changes in density as indicated in Fig.4b the less preferred type might act as a more valuable buffer, to forestall extinction at low prey densities while enabling the population of the preferred type to recover, than if it were harder to catch.

The approach proposed by Rapport (1971) indicates that, even in terms of diet optima for individual survival, complex effects are to be expected once nutrional factors are considered and the idea of complementary foods introduced.

PATCHINESS

The model of Sukhanov and Shapiro (1971) and that developed here each apply to a situation in which the predator is exposed to a variety of prey types at one time. In the sense of MacArthur and Levins (1964) the prey forms a fine-grained resource for the predator. A model of the alternative situation, in which the distribution of the various prey types is course-grained, has been proposed by Royama (1970). Each 'grain' is taken to be occupied by one prey type. The effect of a predator allocating its hunting time between grains in such a way as to maximise the amount of food captured, assuming that within each its rate of predation is given by the 'disc' equation of Holling (1959), (Equation 2 above), is then explored. Clearly insertion of a model of a predator exposed simultaneously to a variety of prey in such a framework would allow heterogeneous prey within each 'grain'.

ACKNOWLEDGEMENTS

I should like to thank Dr.M.B. Usher for his continued advice and encouragement. The paper of Sukhanov and Shapiro was kindly translated for me by Mr.M. Janatka. Much of the work was supported by a grant from the Science Research Council.

SUMMARY

Representation of predation by a predator C upon homogeneous prey F by

$$C + F \xrightarrow{\alpha} CF \xrightarrow{\beta} C$$

in which α and β are rate constants for the changes indicated, enables the equivalent of the most commonly used functional response equation (Holling, 1959; Rashevsky, 1959) to be derived. This approach is extended to incorporate a predator exposed to a number of prey types of equal palatability and generalised to allow preferences between prey. The equations are expressed in terms intended to be easily related to simple experimental observations. Use of the word preference is discussed. The final model is restricted and its implications then explored. The possibility of combining fine and coarse-grain functional response models with heterogeneous prey is raised.

REFERENCES

Emlen, J.M. (1966). The role of time and energy in food preference. *Am. Nat.*, 100, 611-7.

Emlen, J.M. (1967). Optimal choice in animals. *Am. Nat.*, 102, 385-9.

Gause, G.F. (1934). *The struggle for existance.* New York: Hafner Publishing Co.

Haussmann, U.G. (1971). Abstract food webs in ecology. *Mathematical Biosciences*, 11, 291-316.

Holling, C.S. (1959). Some characteristics of simple types of predation and parasitism. *Can. Ent.*, 91, 385-98.

Holling, C.S. (1961). Principles of insect predation. *A. Rev. Ent.*, 6, 163-82.

Holling, C.S. (1965). The functional response of predators to prey density and its role in mimicry and population regulation. *Mem. Ent. Soc. Can.*, 45.

Holling, C.S. (1966). The functional response of invertebrate predators to prey density. *Mem. Ent. Soc. Can.*, 48.

Ivlev, V.S. (1961). *Experimental ecology of the feeding of fishes.* New Haven: Yale University Press.

MacArthur, R. (1955). Fluctuations of animal populations and a measure of community stability. *Ecology*, 36, 533-6.

MacArthur, R. and Levins, R. (1964). Competition, habitat selection, and character displacement in a patchy environment. *Proc. natn. Acad. Sci. U.S.A.*, 51, 1207-10.

MacArthur, R. and Pianka, E.R. (1966). On optimal use of a patchy environment. *Am. Nat.*, 100, 603-9.

Mertz, D.B. and Davies, R.B. (1968). Cannibalism of the pupal stage by adult flour beetles: an experiment and a stochastic model. *Biometrics*, 24, 247-75.

Murdoch, W.W. (1969). Switching in general predators: experiments in predator specificity and stability of prey populations. *Ecol. Monogr.*, 39, 335-54.

Rapport, D.J. (1971). An optimization model of food selection. *Am. Nat.*, 105, 575-87.

Rapport, D.J. and Turner, J.E. (1970). Determination of predator food preferences. *J. theor. Biol.*, 26, 365-72.

Rashevsky, N. (1959). Some remarks on the mathematical theory of nutrition of fishes. *Bull. math. Biophysics*, 21, 161-82.

Rogers, D. (1972). Random search and insect population models. *J. Anim. Ecol.*, 41, 369-83.

Royama, T. (1970). Factors governing the hunting behaviour and selection of food by the great tit (*Parus major* L.).

J. Anim. Ecol., 39, 619-68.
Schoener, T.W. (1969). Models of optimal size for solitary predators. *Am. Nat.*, 103, 277-313.
Solomon, M.E. (1949). The natural control of animal populations. *J. Anim. Ecol.*, 18, 1-35.
Sukhanov, V.V. and Shapiro, A.P. (1971). On selective nutrition of water organisms. *Gidrobiol. Zh. S.S.S.R.*, 7, 63-8. (In Russian).
Watt, K.E.F. (1959). A mathematical model for the effect of densities of attacked and attacking species on the number attacked. *Can. Ent.*, 91, 129-44.
Watt, K.E.F. (1965). Community stability and the strategy of biological control. *Can. Ent.*, 97, 887-95.
Williamson, M.H. (1957). An elementary theory of interspecific competition. *Nature, Lond.*, 180, 422-5.

Switching in invertebrate predators

J.H. LAWTON, J.R. BEDDINGTON and R. BONSER
Department of Biology, University of York

INTRODUCTION

This paper deals with certain aspects of the behaviour of invertebrate predators that are potentially important in influencing the structure and stability of their prey populations. It is therefore concerned with a small facet of the broad set of problems embraced by the term 'ecological stability'.

A considerable body of data supports the idea that predators can have an important effect on the structure of ecological communities (for examples, see Connell, 1971), although by comparison, the development of general, but at the same time reasonable realistic models for populations of polyphagous predators and their prey has been very slow. This is not really very surprising, because the problem is obviously a complicated one which can only be approached by subdivision into a series of relatively discrete component problems. One such problem is concerned with the sorts of prey that predators are likely to take when they are faced with a choice of prey species. In the simplest case, each type of prey may be eaten as some simple function of its relative abundance in the environment; alternatively the predator could show more complicated behaviour by 'switching' - that is by attacking whichever is relatively the most abundant prey supraproportionally (Murdoch, 1969) (For completeness, it should be noted that other sorts of responses are also possible e.g. Allen, 1972; Mueller, 1971).

Existing models come to very different conclusions about

Ecological Stability
edited by M.B. Usher and M.H. Williamson.

the consequences of switching versus non-switching behaviour for the structure of ecological communities (contras
MacArthur, 1972, p.56 with Caswell, 1972 and Slobodkin,
1961) so that the problem requires considerably more investigation. It can be explored further in two ways.
First, it is obviously important to establish whether
switching is, or is not, a normal feature of predator behaviour, and secondly, the consequences of these differences in behaviour for the stability of prey populations
need to be explored both experimentally and in general,
realistic population models. This paper goes a small way
towards both these objectives.

EXPERIMENTS WITH NOTONECTA

Notonecta is a large, actively swimming predator found i
wide variety of freshwater habitats. Adults typically u
four different hunting methods (Ellis and Borden, 1970)
that their behaviour is fairly complicated in comparison
with that shown by many other invertebrates. Because of
this, it was considered possible that they might also
switch.

Adult *Notonecta glauca* L. (Insecta: Hemiptera: Hetero
tera) were collected from ponds near York and held in th
laboratory at 4°C in a large tank containing a wide rang
of prey organisms. Their nutritional history prior to t
experiments was therefore unknown. All experiments were
carried out at room temperature in clear plastic contain
22·5 cm. square, and filled to a depth of 5 cm. with wat
The two prey species were *Asellus aquaticus* L. (Crustace
Isopoda) and larvae of the Mayfly *Cloëon dipterum* (L.)
(Insecta: Ephemeroptera), both of which occur commonly i
the habitats in which *Notonecta* is found. *Asellus* (whic
walk slowly on the substrate and rarely swim) were locat
and captured by *Notonecta* swimming in an inverted manner
along the bottom; Mayfly larvae (which alternate periods
comparative inactivity on the substrate with periods of
rapid swimming) were only occasionally taken by this met
because they normally swim sufficiently rapidly when disturbed to escape. They were usually captured by a previously stationary *Notonecta* darting rapidly down from a
vantage point just below the surface, when they came to
rest close to the predator. Only medium sized *Asellus*
(average length 5mm) and large *Cloëon*, (average length e
cluding the tails 6mm) were used. A total of twenty pre
(of one species or a mixture of both) were renewed daily

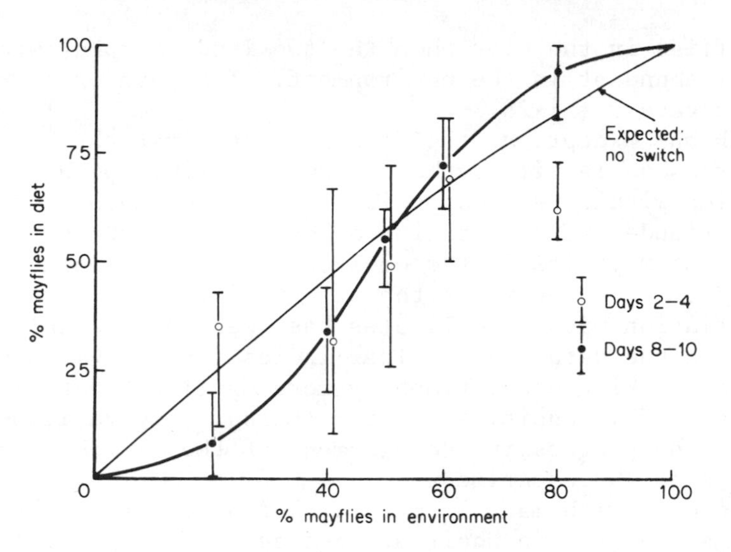

Fig.1. The percentage of mayfly larvae in the diet as a function of their relative abundance in the environment of *Notonecta*, at the start (days 2-4) and end (days 8-10) of the experiment (calculated from the total numbers of prey eaten by each *Notonecta* over each three day period, each point showing the mean and range for five replicates.) The expected proportion of mayflies in the diet was calculated with c = 1·28. *Asellus* were the alternative prey

for each *Notonecta*. Because prey were not replaced as they were eaten, some reduction of prey density occurred during each 24 hour period, although (with the exception of the first day of an experiment when feeding rates tended to be high) the daily variation in prey density was rather small - each *Notonecta* eating an average of 3-4 prey per day.

The first experiment presented *Notonecta* with five different combinations of the two prey (consisting of 20, 40, 50, 60 and 80 percent Mayflies); five replicates of each were used. Fig.1 summarizes the results obtained at the start (days 2-4) and end (days 8-10) of the experiment. The expected proportions of the two prey species in the diet if no switch had occurred were calculated from the equation given by Murdoch (1969)

$$Y = 100\, cX/(100 - x + cX)$$

where X is the percentage prey 1 (Mayflies) in environment, Y is the percentage prey 1 (Mayflies) in diet, and c is a proportionality constant, which can be calculated in a number of ways. In this case c was given by the proportion of Mayflies in the diet when the two kinds of prey were equally abundant in the environment. The mean of 11 replicates gave $c = 1{\cdot}28$.

With one exception, the data for the first three days are a reasonable fit to the 'no switch' line, although the variation within replicates is large. The reason for the apparent under-representation of Mayflies in the diet with 80 percent Mayflies in the environment is not known. At the end of the experiment the situation is very different; the variation within replicates has been considerably reduced and the data show a clear indication of switching - *Notonecta* taking proportionally more Mayflies than expected at high Mayfly densities and vice-versa. Fig.2a illustrates the progressive development of the switch over the ten days of the experiment.

In common with many predators, *Notonecta* are never entirely successful in their attacks and it was possible that the proportion of successful attacks could be improved by learning. This would explain both the steady change in the average performance of animals exposed to different proportions of two different prey (each requiring a different hunting and capture technique) and also the reduction in variation between individual animals, which probably started the experiments with different nutritional histories. A second experiment was therefore carried out in which *Notonecta* were fed for seven days under conditions identical to those used in the switching experiment but with the addition of replicates consisting of 100 percent Mayflies and 100 percent *Asellus*. At the end of this period, the food of the *Notonecta* fed 100 percent Mayflies was replaced with *Asellus* and then all animals in each replicate and treatment were observed continuously until each had been seen to make twenty attacks on *Asellus*. Successful attacks (where the *Asellus* was killed) and unsuccessful attacks (where it escaped) were recorded. (A similar experiment to determine the attack success with Mayflies failed, because unsuccessful attacks on the rapidly swimming Mayflies could not be identified with any degree of certainty.) It is clear from Fig.2b that the proportion of successful attacks on *Asellus* increased as the proportion of *Asellus* in the environment increased. If the probabilities of a predator encountering either of two prey species are simple function of their abundance, but the probability of capturing one (o both) species improves with practice, and if in addition we make some fairly simple assumptions about the rates at whic the predator learns to deal with the prey, and how quickly it 'forgets', then the frequency of a species in the diet o

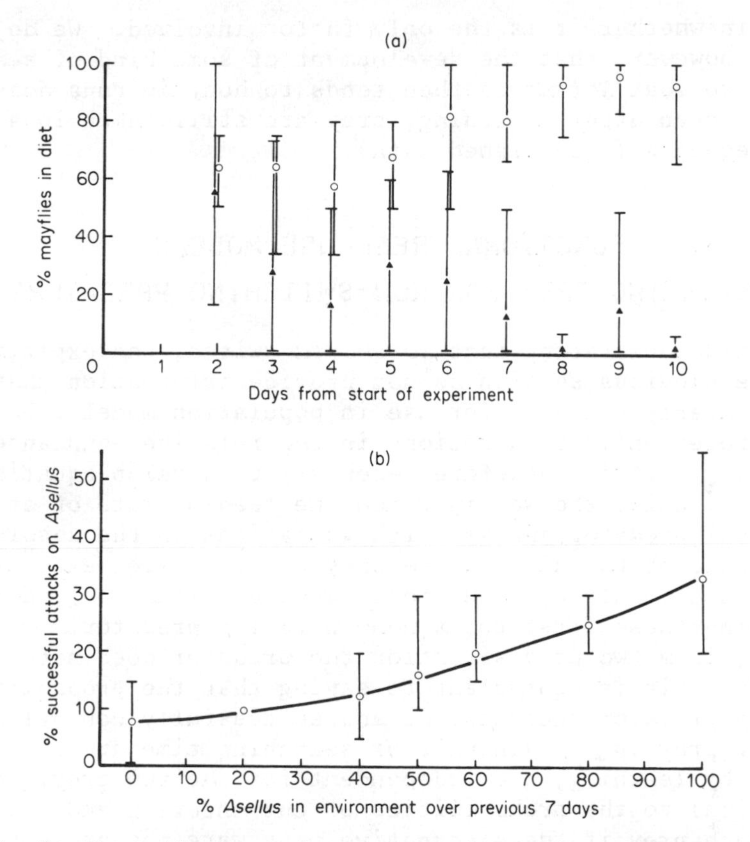

Fig.2. The effect of 'experience' on the prey taken by *Notonecta* (a) shows the development of switching over 10 days (calculated from the prey eaten by *Notonecta* on each day of the experiment, each point showing the mean and range for five replicates). The data for day 1 are omitted because the high feeding rates on the first day reduced prey densities to a significant extent. Open circles represent 80 percent mayflies in the environment, filled triangles 20 percent. (b) shows the proportion of successful attacks by *Notonecta* on *Asellus*, as a function of the proportion of *Asellus* in the environment during the previous seven days (the number of animals, means and ranges are shown; the success rate of all four animals exposed to 20 percent *Asellus* was identical)

a predator will be greater when it is common in the environment than would be predicted from its frequency in the diet when it is rare. This would appear to be a basis for the switch observed in *Notonecta*, although we are not yet certain whether it is the only factor involved. We do know, however, that the development of some kind of search image so that *Notonecta* then tends to hunt in runs does not occur; even after switching, prey are still taken in a random sequence (unpublished data).

FUNCTIONAL RESPONSE MODELS WITH TWO PREY FOR NON-SWITCHING PREDATORS

Whilst demonstrating that *Notonecta* switch, the experiments in the previous section do not provide information that is particularly suitable for use in population models, because they refer only to variations in the relative abundances of two prey. It is therefore necessary to develop equations which describe the way in which the feeding rate of an individual predator changes with variations in the absolute abundance of two alternative prey species (i.e. extending the 'functional response' to a two-prey situation), developing these first for a non-switching predator.

If, in a two-prey situation the predator does not switch, this is equivalent to saying that the probabilities of the predator encountering and successfully capturing either prey in a given unit of searching time are not modified by learning, are independent for the two prey, and identical to the probabilities of encountering and capturing each prey if the alternative prey were not present. If exploitation is unimportant, the appropriate equations are an extension of the 'disc equation' (Holling, 1959). Thus

$$\begin{aligned} E_1 &= a_1 N_1 T/(1 + a_1 h_1 N_1 + a_2 h_2 N_2) \\ E_2 &= a_2 N_2 T/(1 + a_2 h_2 N_2 + a_1 h_1 N_1) \end{aligned} \qquad (1)$$

where the subscripts 1 and 2 refer to the two prey species, E is the total number of prey eaten, N is the total number of prey present (prey density), a is the instantaneous attack-rate constant (the rate of successful search of Holling, 1959), h is the handling time, and T is total time during which the prey are exposed to the predator. (Equations 1 have also been derived by Murdoch, 1973.) In a form suitable for estimating parameters from experimental data by multiple regression these equations become

$$N_1/E_1 = (1/a_1) + h_1N_1 + (a_2h_2N_2/a_1)$$
$$N_2/E_2 = (1/a_2) + h_2N_2 + (a_1h_1N_1/a_2) \quad (2)$$

Equations 1 and 2 are inappropriate for situations where predation makes a significant reduction in prey density and under these circumstances the appropriate equations may be derived by integrating Equations 1, so that we obtain the two-prey equivalent of the 'random predator equation' of Rogers (1972). The great advantage of these equations is that if they are used to estimate parameters from experimental data, it is not necessary to maintain constant densities of prey during the experiments (by adding prey individuals at the same rate as they are removed), in order to estimate the instantaneous coefficients. The equations are

$$E_1 = N_1[1 - \exp\{- a_1 (T - h_1E_1 - h_2E_2)\}]$$
$$E_2 = N_2[1 - \exp\{- a_2 (T - h_2E_2 - h_1E_1)\}] \quad (3)$$

In a form suitable for estimating parameters from experimental data they become

$$\ln S_1 = - a_1T + a_1h_1E_1 + a_1h_2E_2$$

$$\ln S_2 = - a_2T + a_2h_2E_2 + a_2h_1E_1$$

where S is the survival rate of each prey or $(N-E)/N$. The deriviation of these equations is given in the Appendix.

Fig.3 shows the surfaces generated by Equations 1, where the feeding rate on prey 1 alone is higher than on prey 2 alone at comparable densities. Experiments like those carried out on *Notonecta* and by Murdoch (1969) use an experimental design which is essentially a 'slice' across such a surface in the plane of the page. The alternative, which provides information that is more appropriate for population models, is to use a factorial design in which predators are exposed to all combinations of a range of densities of both prey, in order to define the whole surface rather than a segment of it. We do this in the next section.

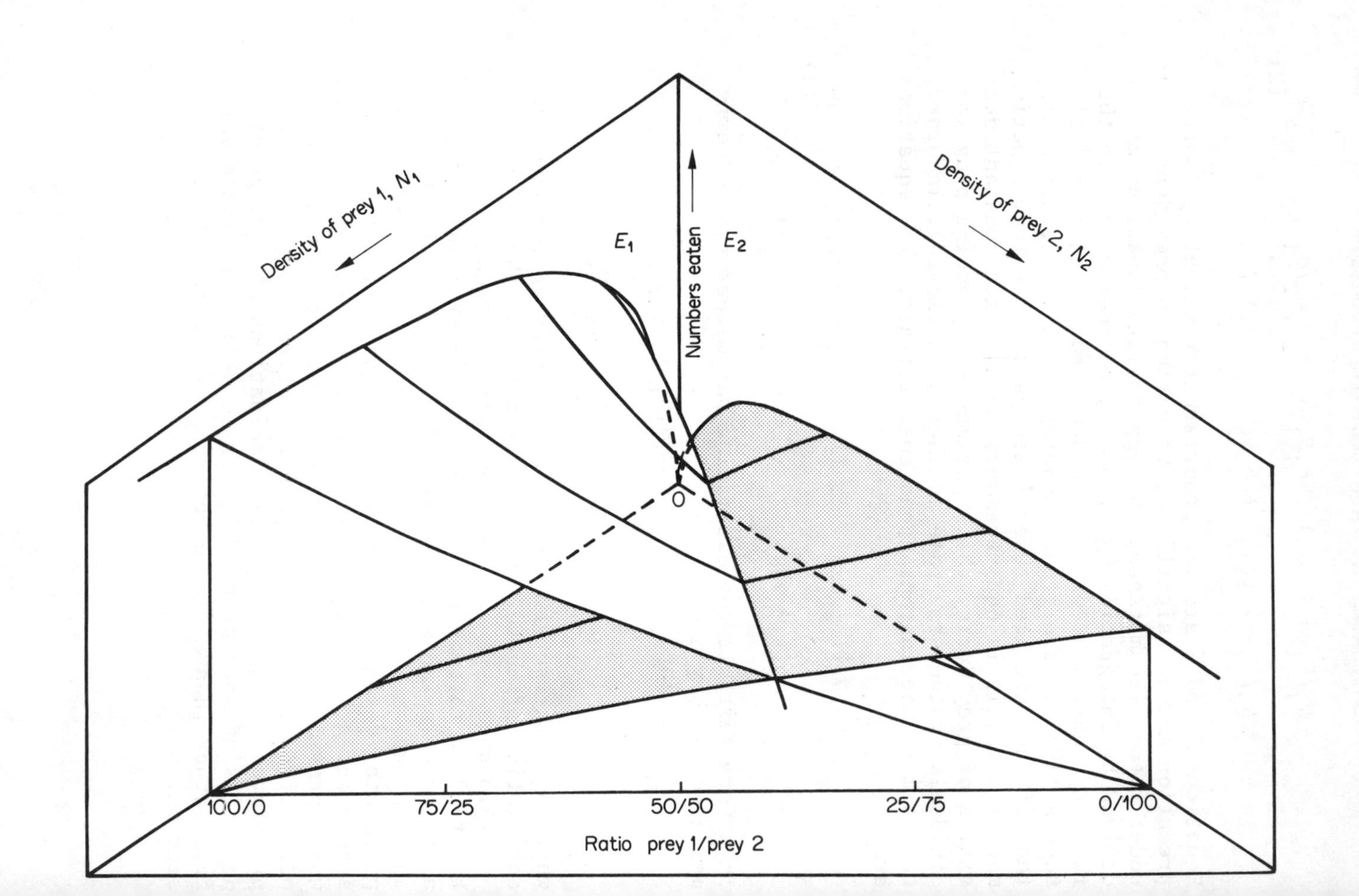

Density of prey 1, N_1
Numbers eaten
E_1
E_2
Density of prey 2, N_2
O
100/0
75/25
50/50
25/75
0/100
Ratio prey 1/prey 2

EXPERIMENTAL ANALYSIS OF PREDATOR RESPONSES TO VARIATIONS IN THE ABSOLUTE DENSITY OF TWO PREY

A factorial experiment, in which predators are exposed to a range of densities of two different prey species can be analysed by the equations developed in the previous section. If these equations provide a good fit to the data then there is no reason to suspect that the predator has switched; a poor fit, however, may be indicative of a more complicated response by the predator, for example switching. We have carried out two such factorial experiments.

The first experiment is essentially an extension of Holling's (1959) 'sandpaper disc' experiment, in which we used a blindfolded human 'predator', searching by tapping with her fingers, for two different sizes of prey (polythene discs) scattered at random over an area of 0•75 m^2. Seven densities of each 'prey' were used (between 0 and 70) in 48 possible combinations. The discs were not replaced once they had been 'predated', but because the amount of exploitation during each run was fairly small, Equations 2 could be employed by using the average abundances of the 'prey' during each run. The results obtained from equations 2 and 4 are summarised in Table 1. Both equations provide satisfactory fits to the experimental data. The tendency for the instantaneous attack-rate constant, a, to be slightly larger when estimated by Equations 4, than when estimated by equations which do not allow for exploitation (Equations 2) has been noted by Rogers (1972) for the analogous single-prey equations. The data are also plotted in Fig.4a, to test for switching in the manner used for *Notonecta*; as expected, the data provide no evidence of switching.

This basic design was also used in a second experiment, except that real predators in the form of penultimate instar larvae of the damselfly *Ischnura elegans* (Lind.) (Odonata: Zygoptera) were used; the larvae were fed two species of Cladocera - *Daphnia obtusa* Kurz and *Simocephalus vetulus* (O.F. Müller). *Simocephalus* differ from *Daphnia* in spending most of their time on the substrate, interspersed by only occasional periods of active swimming. Standard sized prey

---

Fig.3. The predicted functional responses of a predator feeding on two different prey, calculated from Equations 1; in this particular example $a_1 = 1{\cdot}5a_2$, $h_1 = 0{\cdot}04$, $h_2 = 0{\cdot}09$

Table 1. Summary of the results of a factorial experiment with varying densities of two different types of 'prey' being hunted by a human 'predator'; 'prey' 1 were large and 'prey' 2 small polythene discs

	Parameters fitted by Equations 2		Parameters fitted by Equations 4		
Parameter	Simple functional response (N for one or other prey = 0)	With both prey present	Simple functional response (N for one or other prey = 0	With both prey present	
a_1	0·643	0·604	0·668	0·685	
a_2	0·342	0·470	0·347	0·369	
h_1	0·045	0·044	0·047	0·047	0·050
h_2	0·064	0·075	0·064	0·063	0·056
Correlation of multiple correlation coefficients	0·990 (prey 1)	0·852	0·990 (prey 1)	0·803	
	0·950 (prey 2)	0·811	0·960 (prey 2)	0·673	
F values	–	51·71 (d.f. 2,38)	–	35·35 (d.f. 2,39)	
	–	36·56 (d.f. 2,38)	–	16·16 (d.f. 2,39)	

were obtained by passing them through graded sieves, the *Simocephalus* being approximately twice as big as the *Daphnia*. The experiment was carried out at 16°C, in 100ml beakers, each containing a single larva and combinations of both prey species varying between 0 and 75 per beaker. The larvae were starved for seven hours prior to the experiment and were then allowed to feed for sixteen hours. Table 2 and Figs.4b and 5 summarize the results.

It is clear from Table 2 that (unlike the polythene discs experiment) the fit of the experimental data to Equations 4 was very poor. In part this poor fit may be attributed to the large amount of individual variation in feeding rates shown by the larvae; but in addition, Fig.4b

Table 2. Summary of the results of a factorial experiment with varying densities of two different types of prey (*Daphnia obtusa* and *Simocephalus vetulus*) being fed on by larvae of the damselfly *Ischnura elegans*; parameters were estimated using Equations 4. Prey 1 refers to *Daphnia* and prey 2 to *Simocephalus*

Parameter	Simple functional response (N for one or other prey = 0)	With both prey present	
a_1	1·662	0·747	
a_2	0·498	0·217	
h_1	0·034	0·0005	0·018
h_2	0·064	0·019	0·125
Correlation or multiple correlation coefficients	0·823 (prey 1)	0·132	
	0·616 (prey 2)	0·323	
F values	-	0·327 (d.f. 2,37)	
	-	2·159 (d.f. 2,37)	

shows an indication of the larvae having switched. Two alternative estimates of the no-switch line are shown in Fig.4b, based on two different estimates of c; one was obtained by the regression of E_1/E_2 against N_1/N_2 where the

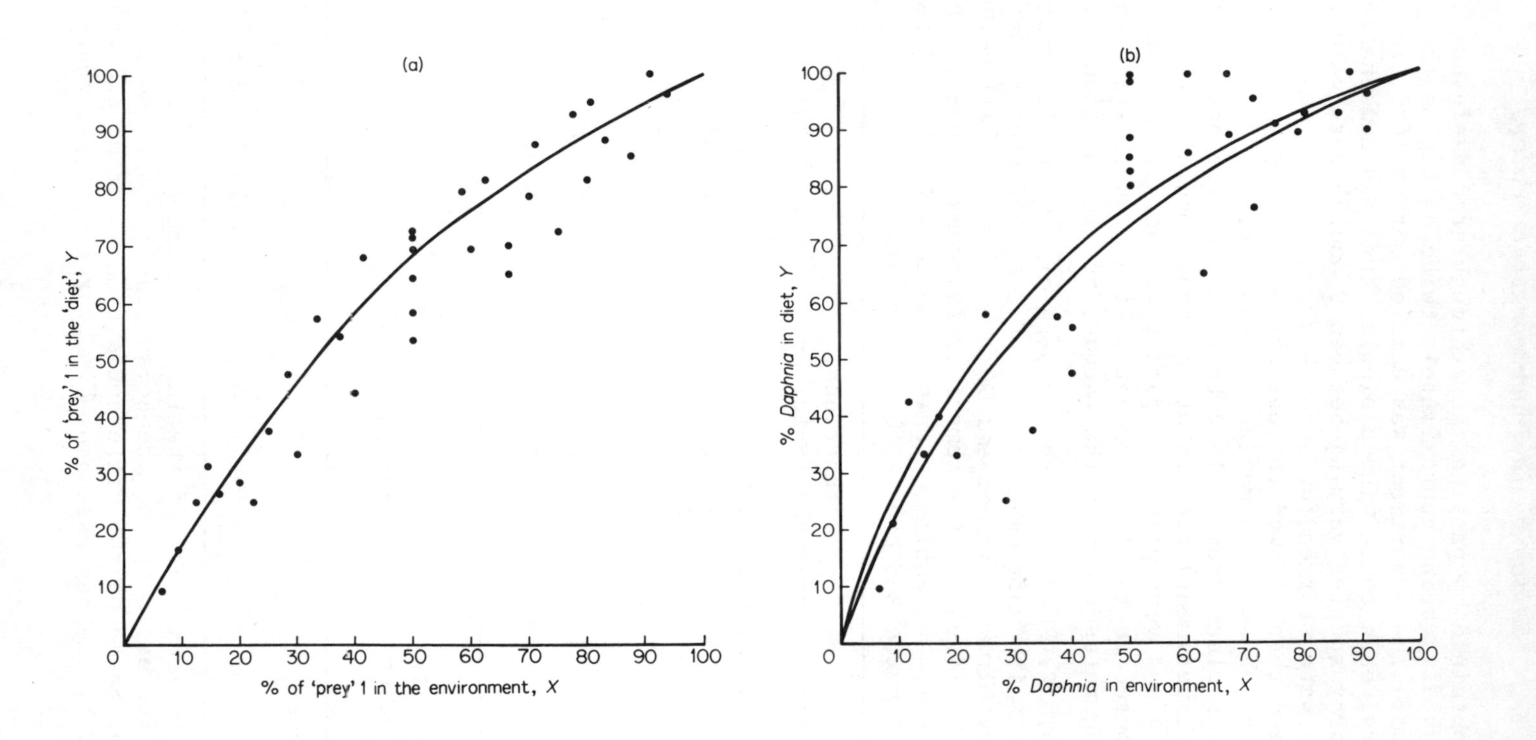
(a)
% of 'prey' 1 in the 'diet', Y
% of 'prey' 1 in the environment, X
(b)
% Daphnia in diet, Y
% Daphnia in environment, X

line is forced through the origin, giving $c = 2{\cdot}70$; and the other was obtained by dividing Equations 1 for species 1 by the same equation for species 2. Where $N_1 = N_2$, we then obtain, after cancelling, $c = a_1/a_2 = 3{\cdot}34$. Where a switch is suspected, this second estimate of c is theoretically more satisfactory than the first. A third estimate of c, which was used with the *Notonecta* data is the ratio E_1/E_2 when $N_1 = N_2$. In the present experiment, the actual data appear to cross the 'no switch' line at a point below that where $N_1 = N_2$ so that the ratio of E_1/E_2 at $N_1 = N_2$ is not an appropriate estimate of c. (In the polythene discs experiment, all three methods of estimating c give more or less identical results.)

Whilst this experiment with *Ishnura* appears to provide a further example of switching by an invertebrate predator, the reason for the switch is not yet clear; a possible mechanism is discussed in the next section.

MODIFICATION OF THE FUNCTIONAL RESPONSE MODELS TO INCLUDE SWITCHING

The response of a predator to changes in the absolute density of two prey species may be more complicated than that predicted by Equations 1 and 3 because the probability of the predator encountering and successfully capturing one species in a given unit of searching time does not depend on just the density of that one species but is in some way modified by the fact that two prey species are present. Two broad categories of possibilities are:

(1) The probability of successfully *capturing* a prey in-

Fig.4. Results of two factorial experiments in which predators were exposed to combinations of densities of two prey. In (a) the 'predator' was a blindfolded human being searching for two kinds of 'prey' (large and small polythene discs); no switch occurred. The curve shows the expected proportion of the two prey with no switch, $c = 1{\cdot}95$. In (b) the predators were larvae of the damselfly *Ischnura* feeding on *Daphnia* and *Simocephalus*; the data suggest that a switch may have occurred. The lines show the expected proportion of the two prey with no switch and $c = 3{\cdot}34$ (upper line) and $c = 2{\cdot}70$ (lower line)

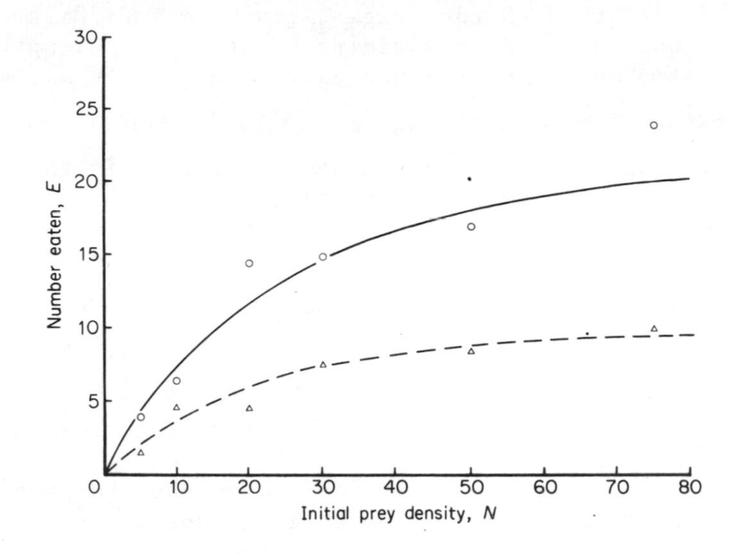

Fig.5. The functional responses of *Ischnura* with *Daphnia* (solid line) and *Simocephalus* (dashed line) each in the absence of the second prey species. a = 1•662 and h = •034 for *Daphnia* and a = 0•498 and h = •064 for *Simocephalus*

creases as some function of the rate at which the predator encounters it; alternatively, the effect of encountering individuals of a second species (which require a different technique to capture them) may make the predator less successful in dealing with the first species (the Jack of all trades, master of none principle). These effects could work separately or together. Something akin to this appears to have been operating with *Notonecta*, and may apply generally to any predator which feeds on a variety of prey, involving several different methods of capture, where the prey are active and likely to escape and where the predators' performance may be improved by learning. The instantaneous attack-rates in Equation 1 may be modified in a number of ways to include effects of this kind; these will be reported in a later publication when the effect of learning on the instantaneous attack rate of *Notonecta* has been explored more fully.

(2) The probability of *encountering* a prey species changes, either because one species becomes more difficult or easier to find in the presence of the other (for example because of interactions between the prey) or because of some change in the internal state of the predator. This could involve the formation of some kind of 'search image'

where the predator 'learns to see' the prey (Dawkins, 1971) or it could involve changes in the instantaneous attack-rate through the hunger-level of the predator (Holling, 1966). Relatively short-term experiments like those with *Ischnura* may involve continuous changes in the hunger-level of the predator during the experiment, so that estimates of the instantaneous attack rate on a single prey species are strictly only applicable under the average hunger levels prevailing during the experiment. If we assume that consumption of one individual of the two different prey species changes the hunger-level of the predator at different absolute rates, then the probability of the predator responding to, say, a small rare prey may then be less in the presence of a large numerous prey (capture of which reduces the level of hunger substantially) than when the small prey is the only food available. It is possible that this sort of effect was *partly* responsible for the switch observed with *Ischnura*, and explains why fewer *Daphnia* than expected were taken when the larger *Simocephalus* was relatively more abundant. An exploration of the effects of hunger on the response thresholds of predators to different prey, using models rather different to ours, is reported by Harris in this volume. The long term effects of hunger on switching, however, are unclear, although it seems probable that they are likely to be less important than learning.

CONCLUSIONS

Whilst there is little doubt that birds and mammals have the necessary behavioural attributes to show switching (e.g. Alcock, 1972; Allen, 1973; Clarke, 1969;Manly *et al.*, 1972) evidence for similar behaviour in invertebrate predators has been very meagre (Murdoch, 1969; Murdoch and Marks, 1973; but see Rapport, 1971; Richman and Rogers, 1969). It now appears that switching may occur more generally than has previously been suspected in invertebrate predators and for a variety of different reasons (see also Murdoch and Marks, 1973) so that further exploration of the consequences of this behaviour for prey populations appears to be justified.

We have said nothing in this paper about how invertebrate predators might be expected to respond to patches of prey (Hassell and Rogers, 1972; Royama, 1971), our analysis being concerned entirely with what happens 'within patches'. This and other aspects of the population dynamics and life-histories of the predator and its prey species are separate problems each requiring further investigation before a more detailed understanding of the stability of polyphagous predators and their prey populations is possible.

ACKNOWLEDGEMENTS

We are extremely grateful to Mrs. Ann Fisher for technical help and to several colleagues at York, Imperial College Field Station and Oxford for their helpful comments.

SUMMARY

Switching by polyphagous predators is a potentially important stabilising mechanism. Switching by two species of aquatic invertebrate predators, *Notonecta* and *Ischnura*, is reported under laboratory conditions. We develop the functional response equations for a single species of prey so that they can be applied to situations involving more than one species. Methods of fitting these equations to experimental data are described.

REFERENCES

Alcock, J. (1973). Clues used in searching for food by red-winged blackbirds (*Agelaius phoeniceus*). *Behaviour*, 46, 174-88.

Allen, J.A. (1972). Evidence for stabilizing and apostatic selection by wild blackbirds. *Nature*, Lond., 237, 348-9.

Caswell, H. (1972). A simulation study of a time lag population model. *J. theor. Biol.*, 34, 419-39.

Clarke, B. (1969). The evidence for apostatic selection. *Heredity*, 24, 347-52.

Connell, J.H. (1971). On the role of natural enemies in preventing competitive exclusion in some marine animals and in rain forest trees. *Dynamics of Populations* (Ed. by P.J. den Boer and G.R. Gradwell), pp.298-312. Wageningen: PUDOC.

Dawkins, M. (1971). Perceptual changes in chicks: another look at the 'search image' concept. *Anim. Behav.*, 19, 566-74.

Ellis, R.A. and Borden, J.H. (1970). Predation by *Notonecta undulata* (Heteroptera: Notonectidae) on larvae of the yellow-fever mosquito. *Ann. ent. Soc. Am.*, 63, 963-73.

Hassell, M.P. and Rogers, D.J. (1972). Insect parasite responses in the development of population models. *J. Anim. Ecol.*, 41, 661-76.

Holling, C.S. (1959). Some characteristics of simple types of predation and parasitism. *Can. Ent.*, 91, 385-98.

Holling, C.S. (1966). The functional response of invertebrate predators to prey density. *Mem. Ent. Soc. Can.*, 48.

MacArthur, R.H. (1972). *Geographical Ecology.* New York: Harper and Row.

Manly, B.F.J., Miller, P. and Cook, L.M. (1972). Analysis of a selective predation experiment. *Am. Nat.*, 106, 719-36.

Mueller, H.C. (1971). Oddity and specific searching image more important than conspicuousness in prey selection. *Nature, Lond.*, 233, 345-6.

Murdoch, W.W. (1969). Switching in general predators: experiments on predator specificity and stability of prey populations. *Ecol. Monogr.*, 39, 335-54.

Murdoch, W.W. (1973). The functional response of predators. *J. appl. Ecol.*, 10, 335-42.

Murdoch, W.W. and Marks, J.R. (1973). Predation by coccinellid beetles: experiments on switching. *Ecology*, 54, 160-7.

Rapport, D.J. (1971). An optimization model of food selection. *Am. Nat.*, 105, 575-87.

Richman, S. and Rogers, J.N. (1969). The feeding of *Calanus heligolandicus* on synchronously growing populations of the marine diatom *Ditylum brightwellii*. *Limnol. Oceanogr.*, 14, 701-9.

Rogers, D. (1972). Random search and insect population models. *J. Anim. Ecol.*, 41, 369-83.

Royama, T. (1971). Evolutionary significance of predators' response to local differences in prey density: a theoretical study. *Dynamics of Populations* (Ed. by P.J. den Boer and G.R. Gradwell), pp.344-57. Wageningen: PUDOC.

Slobodkin, L.B. (1961). *Growth and Regulation of Animal Populations.* New York: Holt, Rinehart and Winston.

APPENDIX

We have from Equations 1 that the instantaneous attack rates are given by

$$E_1 = \frac{a_1 N_1}{1 + a_1 h_1 N_1 + a_2 h_2 N_2}$$

$$E_2 = \frac{a_2 N_2}{1 + a_1 h_1 N_1 + a_2 h_2 N_2}$$

These can be considered as instantaneous death rates of the two species: thus we have

$$\frac{dN_1}{dt} = \frac{-a_1 N_1}{1 + a_1 h_1 N_1 + a_2 h_2 N_2} \qquad (A1)$$

$$\frac{dN_2}{dt} = \frac{-a_2 N_2}{1 + a_1 h_1 N_1 + a_2 h_2 N_2} \qquad (A2)$$

From Equation A1 we have

$$\frac{(1 + a_1 h_1 N_1 + a_2 h_2 N_2)\, dN_1}{N_1} = -a_1 dt \qquad (A3)$$

rearranging and making the substitution $N_2/N_1 = a_1 dN_2/a_2 dN_1$ we have

$$\int_{N_1}^{N_1 - E_1} (1/N_1 + a_1 h_1) dN_1 + \int_{N_2}^{N_2 - E_2} a_1 h_2 dN_2 = \int_0^T - a_1 dt \qquad (A4)$$

This integrates to give $\{(N_1 - E_1)/N_1\}$

$$\log \{(N_1 - E_1)/N_1\} - a_1 h_1 E_1 - a_1 h_2 E_2 = -a_1 T \qquad (A5)$$

which is the form used in Equations 4 with S_1 defined as $(N_1 - E_1)/N_1$. The corresponding equation for the second species can obviously be derived in an analogous manner. Equation A5 can be rearranged to yield a form similar to that of Rogers (1972) for the single species case. This is

$$E_1 = N_1 \left[1 - \exp \{-a_1 (T - h_1 E_1 - h_2 E_2)\}\right] \qquad (A6)$$

The integration of theory and experiment in the study of predator–prey dynamics

M. J. BAZIN, V. RAPA and P. T. SAUNDERS
Queen Elizabeth College, University of London

It has been known for a long time (D'Ancona, 1926) that when the populations of two species are principally controlled by one of the species preying upon, and consequently being dependent upon, the other, both populations tend to fluctuate. While the maxima of the predator population occur somewhat after those of the prey, the intervals between maxima are approximately the same for both populations. Lotka (1920) and Volterra (1931) independently proposed the so-called Lotka-Volterra equations

$$dN_1/dt = a_1N_1 - b_1N_1N_2$$

$$dN_2/dt = -a_2N_2 + b_2N_1N_2$$

which appear to be the simplest plausible model for such prey-predator interactions. These equations do not have known analytical solutions, but it has nevertheless been shown by Goel *et al.* (1971) that the solutions are at least qualitatively correct in that they are all oscillatory, with N_1 and N_2 having the same period but out of phase.

While the Lotka-Volterra equations do produce the correct general behaviour, we have been unable to find any report of data from a natural ecosystem which appears to give a good fit to the model. Further, Goel *et al.* (1971) and Samuelson (1971) have shown that either random fluctuations in the parameters or a law of diminishing returns (e.g., a Verhulst term) in the expression for the growth of the prey in the absence of predator will lead to a damping of the oscilla-

Ecological Stability
edited by M.B. Usher and M.H. Williamson.

tions. As Samuelson pointed out, if we accept this model it is surprising that oscillations are observed in nature at all, unless there are relatively frequent large scale perturbations such as changes in climate.

We have thus on the one hand the very simple mathematical model of Lotka and Volterra, and on the other the complicated ecological systems found in nature. The former is evidently insufficient to describe the latter, although it is clearly of great value if only as an aid to understanding the general nature of the interaction. But if we want to do better than this we must attempt to bridge the gap that exists between the model and nature, and there are obviously two ways of going about this: either we must make the model more complicated or else we must study particularly simple ecosystems. We expect, of course, that it is only through a combination of both approaches that the problem will eventually be solved. For reasons of space we shall not attempt to discuss the valuable mathematical work which has been done in this field; see, for example, the review of Goel *et al.* (1971). We shall confine ourselves to the experimental study of comparatively simple prey-predator systems studied by Tsuchiya *et al.* (1972), Canale (1970), Curds and Cockburn (1968) and ourselves.

In the microbial ecosystem that we are studying, we are using amoebae of the cellular slime mould *Dictyostelium discoideum* feeding on *Escherichia coli*. The organisms are grown in chemostat culture, so the system is open with respect to nutrient and energy flow as it would be in nature. The effect of the abiotic environment, represented here by the nutrient (glucose) limiting bacterial growth, can be precisely measured and can be taken into account quantitatively by considering the nutrient to be a third 'species' in the generalised Lotka-Volterra equations (Goel *et al.*, 1971)

$$\mathrm{d}N_i/\mathrm{d}t = a_i N_i - \sum_j b_{ij} N_i N_j$$

$$b_{ii} = 0$$

The experimental parameters (i.e. the dilution rate and the limiting nutrient concentration in the input medium) are easily controlled and enter into the equations in a straightforward fashion. The number of individuals of each species is very large and the system is well mixed so statistical fluctuations and sampling errors should be negligible. The environment can be controlled so that random perturbations are minimised, while the short generation time of the organisms means that experiments can be conducted over a period of a few days, rather than the months required for experiments with insects or the years needed for observations of animals in the wild; for the same reasons experiments can be repeated, either with or without changes in conditions.

The work reported here is still in progress and we are therefore not in a position to give final results and conclusions, but even at this stage there are three particular points on which we are able to comment:

Population and Biomass

The variables N_i in the Lotka-Volterra equations are taken to be the prey and predator densities, but in comparing our results with theoretical predictions we must specify whether this density is measured in terms of numbers or biomass. Since biomass may be estimated by taking the product of the mean cell volume and the number density, there would be no difficulty if the mean cell volume was constant. We have measured both the mean cell volume and the number density of the predator, *Dictyostelium discoideum*, and have found that not only does the mean cell volume not remain constant, but that it oscillates and that its oscillations are out of phase with those of the number density. We shall continue to measure both values. Should it turn out that a significantly better agreement with theory can be obtained by using one measure of density than the other, this might give some insight into the mechanism of predation or of growth in general.

The Specific Feeding Rate

In their simplest form, the Lotka-Volterra equations assume that the specific feeding rate of the predator (i.e. $\mathring{P}/P$ in the absence of death) is a constant multiple of the prey density. Other relationships have also been supposed; in particular, for microbial systems, the saturation kinetics proposed by Monod (1942) have been used to describe the specific rate of feeding (Canale, 1970). The results of Curds and Cockburn (1968), who used data from a batch culture and ourselves, with formalin-killed and stationary phase bacteria, show that the specific feeding rate reaches a maximum and then actually declines if prey density is further increased, so, as is shown in Fig.1, neither of the theoretical curves fits the experimental data, but of the two the saturation model gives the better approximation. A better fit might be obtained if the Monod term is modified to take account of substrate (here prey) inhibition.

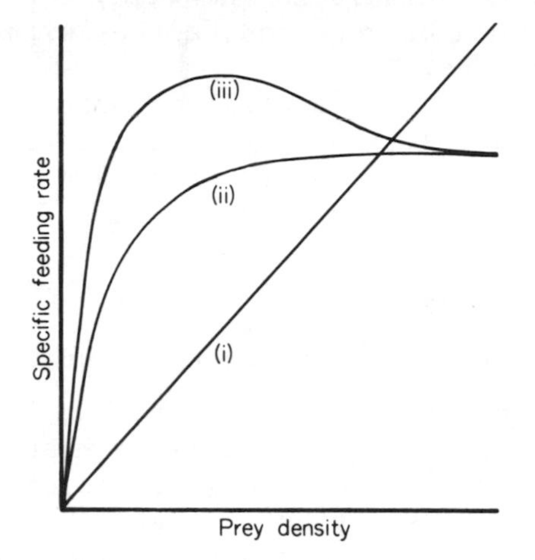

Fig.1. Specific rate of feeding as predicted by the Lotka-Volterra equations (i) and the Monod saturation function (ii) compared with a synthesis of experimental results (iii)

Stability

The oscillations predicted by the Lotka-Volterra equations are unstable, in the sense that almost any realistic perturbation of the system will produce damping. If we plot N_1 and N_2 in a phase plane diagram, then the solutions of the Lotka-Volterra equations are closed trajectories about the equilibrium point, which is a vortex, while if we perturb the system (by adding a Verhulst term, for instance) the equilibrium point becomes a stable focus and the trajectories spiral in towards it. It can be shown, Canale (1970), that if a saturation term of the Monod type is introduced and we then plot substrate, prey and predator in a phase space diagram, the trajectories, which in general will be twisted curves, all tend towards a certain plane, and that subject to certain conditions on the parameters, they tend towards a limit cycle in this plane. This implies that the oscillations are stable, i.e. if the amplitude is either increased or decreased by some small perturbation it will return to the unique value fixed by the conditions of the experiment but independent of the initial populations.

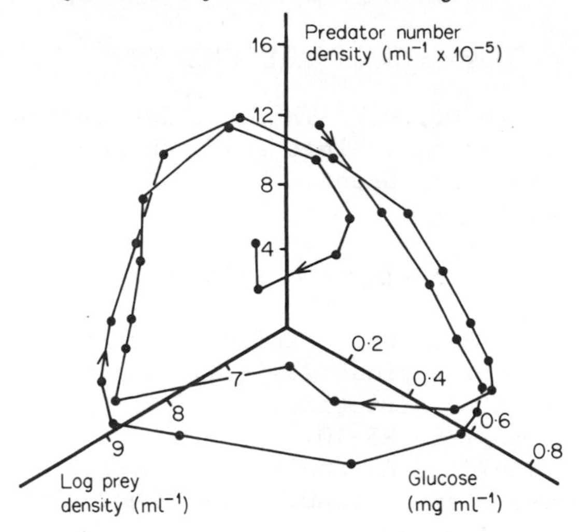

Fig.2. Phase space plot of results from a chemostat culture of *Dictyostelium discoideum* feeding on *Escherichia coli* with glucose as the limiting substrate

Fig.2 is a phase space diagram of some of our results. The data are consistent with the hypothesis that the trajectory should tend towards a plane, but it is not clear whether it is tending towards a point or a limit cycle. Experiments are now in progress to distinguish between the two possibilities; this is being done by running the system under the same environmental conditions but with initial populations as close as possible to the equilibrium point indicated by the original experiment. If in fact a limit cycle exists, then the amplitude of the oscillations should increase. Such a result would support the hypothesis that a saturation term should be included in the governing equations; it would also help us to explain the observed persistence of oscillations without recourse to the large scale perturbations suggested by Samuelson (1971), and might suggest that similar effects may occur in other systems.

We believe that even our preliminary results illustrate the considerable advantages to be gained from the use of microorganisms in the study of ecological systems. We are able to construct and test hypotheses as in any other experimental science, rather than being forced to make plausible deductions from passive observations. We are confident that continued work along these lines by both theoreticians and experimenters will enable us fully to understand the relatively simple interactions of microbes, and thus provide a sound basis for an eventual theory which will explain the more complicated ecosystems found in nature.

ACKNOWLEDGEMENTS

We are grateful to Dr. C.F. Thurston for many helpful and stimulating discussions. This research is being supported by a N.E.R.C. Research Grant.

REFERENCES

D'Ancona, U. (1926). Dell 'influenza della stasi peschereccia del periodo 1914-1918 sul patrimonico ittico dell' Alto Adriatico. *R. Comitato Talass. Italiano. Mem.,* 126, 95-107.

Canale, R.P. (1970). An analysis of models describing predator-prey interactions. *Biotechnol. Bioeng.,* 12, 353-78.

Curds, C.R. and Cockburn, A. (1968). Studies on the growth and feeding of *Tetrahymena pyriformis* in axenic and monaxenic culture. *J. gen. Microbiol.,* 54, 343-58.

Goel, N.S., Maitra, S.C. and Montroll, E.C. (1971). On the Volterra and other non linear models of interacting populations. *Rev. Mod. Phys.,* 43, 231-75.

Lotka, A.J. (1920). Analytical note on certain rythmic relations in organic systems. *Proc. natn. Acad. Sci. U.S.A.,* 6, 410-5.

Monod, J. (1942). *Récherches sur la croissance des cultures bacteriennes.* Paris: Hermann.

Samuelson, P.A. (1971). Generalised predator-prey oscillations in ecological and economic equilibrium. *Proc. natn. Acad. Sci. U.S.A.,* 68, 980-3.

Tsuchiya, H.M., Drake, J.F., Jost, J.L. and Fredrickson, A.G. (1972). Predator-prey interactions of *Dictyostelium discoideum* and *Escherichia coli* in continuous culture. *J. Bact.,* 110, 1147-53.

Volterra, V. (1931). Lecons sur la théorie mathématique de la lutte pour la vie. Paris: Gauthier-Villars.

Food web linkage complexity and stability in a model ecosystem

M. I. WEBBER

Edward Grey Institute of Field Ornithology,
Department of Zoology, University of Oxford.

INTRODUCTION

Elton (1958) gave as one reason why more complex ecosystems might be expected to be more stable than simpler ones, that simple, two species, ecosystem models tend to be unstable. As May (1971, 1973), has pointed out, such an argument is only valid if it may be shown that more complex ecosystem models, involving more species with a greater degree of food web linkage, exhibit greater stability than simpler ones. Recent studies of ecosystem models (May, 1971, 1973; Gardner and Ashby, 1970), have suggested that this is not the case, in particular May concluded 'that if we contrast simple few species mathematical models with the analogously simple multispecies models, the latter are in general less stable than the former'. In this paper, the question of stability is examined in a multispecies difference equation model under two contrasting degrees of linkage complexity, in a randomly fluctuating environment.

THE MODEL

The structure of the model ecosystem is shown in Fig.1a. It consists of three trophic layers with 2, 4 and 8 populations in each layer respectively. Each population is governed by a difference equation describing its dynamics between successive generations, in the form

---

Ecological Stability
edited by M.B. Usher and M.H. Williamson.

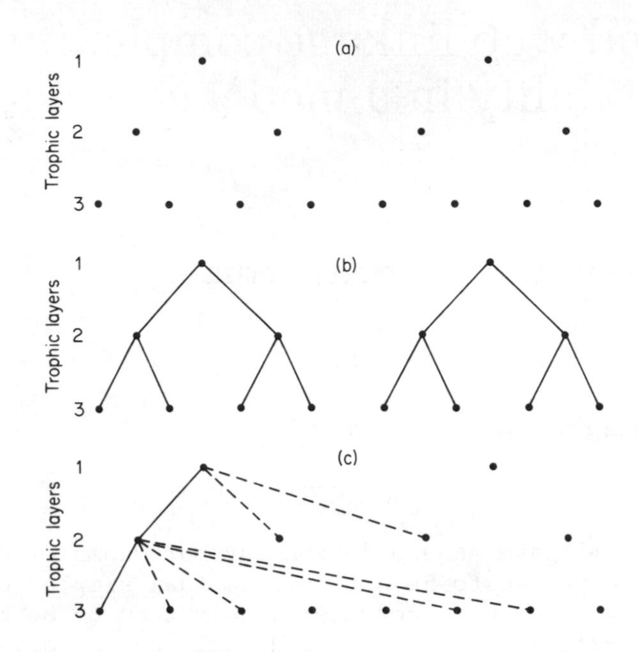

Fig.1. The trophic structure of the model ecosystem (a). (b) The simple food web linkage, in which the solid lines indicate predator-prey relations. (c) The complex food web linkage, showing links for populations 1,1 and 2,1 only. Solid lines indicate 'food refuge' predator-prey relations, and dashed lines indicate competitive predator-prey interactions

$$\log N_{x,j,n+1} = \log N_{x,j,n} + R_x(1 - K_{x,j,n}) - C_{x,j,n} + \log S \quad (1)$$

where, in the subscripts,

x refers to the trophic layer,

j to a population in the xth trophic layer, and

n is the generation number, i.e. time in generations,

and

$N_{x,j,n}$ is the size of population x,j in generation n,

R_x is the intrinsic rate of natural increase for all populations in x,

$K_{x,j,n}$ and $C_{x,j,n}$ are the carrying capacity and predation functions defined in Equations 2 and 3 respectively,

S is the survival rate parameter, the probability of survival so that $0 \leq S \leq 1$.

Equation 1 is derived from the predator prey differential equations formulated by Leslie (1948, 1958), extended for a multiple trophic layer system and differenced in the logarithmic form (Williamson, 1972, p.132). It may be regarded as an extension of the logistic, consisting of a series of parts describing the reproductive rate, density dependent mortality, predation mortality, and non-density dependent mortality.

Two contrasting linkage schemes are used to connect populations in the model. In the 'simple' scheme, each species in trophic layers 1 and 2 preys on two species in the trophic layer directly below it, each species in trophic layers 2 and 3 being preyed on by a single predatory species, as shown in Fig.1b. Thus there is no competition between predator populations for prey. The system splits into two identical parts. In the 'complex' scheme, each species in trophic layers 1 and 2 preys on one species in the trophic layer directly below it for which it is the only predator, this species forming its 'food refuge' (Reynoldson and Davies, 1970). The remaining species in trophic layers 2 and 3 are preyed on by all the species in the trophic layer above. The linkages for one species in trophic layer 1 and one species in trophic layer 2 are shown in Fig.1c.

The linkage schemes are incorporated into the model through the equations for the functions K and C, which take the form

$$K_{x,j,n} \begin{cases} = A_x \sum_k Q_{x,j\to k} N_{x+1,k,n}, & \text{for trophic layers 1,2} \\ = \text{a constant}, & \text{for trophic layer 3} \end{cases} \quad (2)$$

$$C_{x,j,n} \begin{cases} = B_x \sum_i P_{x,j\leftarrow i} N_{x-1,i,n}, & \text{for trophic layers 2,3} \\ = 0, & \text{for trophic layer 1} \end{cases} \quad (3)$$

where, in the subscripts,

i is a population in trophic layer $x + 1$,
j is a population in trophic layer x,

k is a population in trophic layer $x - 1$,
n is the generation number,

and

A_x, B_x are constants, specific to trophic layer x,
N is as in Equation 1,
$P_{x,j \leftarrow i}$ is the 'predation constant' for population x,j being preyed on by population $x - 1,i$, i.e. the proportion of the predation of population x,j due to population $x - 1,i$ when all populations in the trophic layer $x - 1$ are equally abundant,
$Q_{x,j \rightarrow k}$ is the 'food preference constant' for population x,j on population $x + 1,k$, i.e. the proportion of the food of population x,j taken from population $x + 1,k$ when all populations in trophic layer $x + 1$ are equally abundant.
P and Q are constrained so that:

$$\sum_k Q_{x,j \rightarrow k} = \sum_i P_{x,j \leftarrow i} = 1$$

In the complex linkage scheme the amount of any prey species taken by a predator population is not affected by the amounts taken by other predators within a single generation, i.e. there is no direct interference, but the total amount of predation on any prey population influences its abundance in the succeeding generation and so the predators are in competition between generations.

The value of $P_{x,j \leftarrow i}$ and $Q_{x,j \rightarrow k}$ for the two linkage schemes are given in Appendix 1.

NEIGHBOURHOOD STABILITY ANALYSIS

To perform a neighbourhood stability analysis of the difference equation system the values of the parameters in Equations 1, 2 and 3 were constrained so that the equilibrium levels of all the populations in trophic layers 1, 2 and 3 were 10, 100 and 1,000 respectively, and a value of 0·7 assigned to S, the survival parameter in Equation 1. These constraints allowed the values of R_x and B_x determining the reproductive rates and percentage mortality due to predation, to be independently varied, so that for any given set of values of R_x and B_x the remaining parameters in the equations were entirely specified.

May (1973) has shown that for a set of difference equations describing the dynamics of a community of m species

$$N_{j,t-1} - N_{j,t} = F_j(N_{1,t}, N_{2,t}, \dots, N_{m,t}) \tag{4}$$

where $N_{j,t}$ is the size of population j at time t, the neighbourhood stability is described by the eigenvalues of the matrix A the coefficients of which, $a_{j,k}$, measure the effect of the kth species upon the jth and are derived from the function

$$a_{j,k} = \frac{N_{k*}}{N_{j*}} \left(\frac{\partial F_j}{\partial N_k} \right)^* \tag{5}$$

where the term in brackets is the partial derivative of F_j with respect to N_k, evaluated at the equilibrium; F_j is the homologous differential equation to the difference equation F_j; and N_{k*} is the equilibrium level of population k.

The equilibrium interaction matrix A was thus constructed for the model difference equation system, for both of the linkage schemes. The coefficients have the correct signs to fulfil the conditions of qualitative stability outlined by May (1973). The eigenvalues for the two matrices were evaluated for a number of different sets of values for R_x and B_x. May (1973) has shown that the set of difference equations F_j will be stable in the neighbourhood of the equilibrium, if and only if

$$|\lambda_j + 1| < 1 \text{ for all } j \tag{6}$$

where λ_j is the jth eigenvalue of matrix A. A study of the largest values of $|\lambda_j + 1|$ for sets of values of R_x and B_x for the two linkage schemes gives rise to the following conclusions

(1) Both linkage schemes are only quantitatively stable for the two lower value sets of R_x at the lowest value set of B_x.

(2) Whilst the largest values of $|\lambda_j + 1|$ vary between the two linkage schemes, sometimes quite markedly, there is no consistent difference between them.

(3) Stability decreases with increased values of B_x and at high and some low values of R_x.

COMPUTER SIMULATION

A computer simulation of the model was performed, using the parameter values outlined in Table 1, to explore the stability of the two linkage schemes under conditions of a

Table 1. Values of R_x and B_x used in computer simulation

x	1	2	3
R_x	0•5	0•8	1•0
B_x	0•0	0•02	0•002

fluctuating environment. Random environmental variations were introduced through the parameter S in Equation 1. Values for S were produced by a normally distributed pseudo-random number generator, around a mean value of 0•7 and standard deviation s value of 0•05. All the other parameters remained deterministic.

The stability of the two linkage schemes was measured by running the simulation for 100 generations and estimating the standard deviation of the logarithm of the size of each population x,j. Williamson (1972) discusses the use of the standard deviation of the log population size as an ecological measure of variability and stability and Watt (1965) has used the standard error of log population size as a measure of stability for data on certain Canadian forest insect pest species. The advantages of using the standard deviation as a measure of population stability are that, together with the log population mean, it provides a measure relatable to the probability that the population will drop below a specified level, and thus it may be used to describe field populations, being readily calculable from field data. Describing the stability of the ecosystem model in terms of the standard deviations of its component species is less satisfactory than analytically exploring the global stability properties of the difference equations, under stochastic conditions, but the set of difference equations comprising the model are not readily amenable to such analysis.

Analogous to May's (1973) criteria for semi-stochastic

differential equation systems, the condition for neighbourhood stability becomes

$$|\lambda_j + 1| < 1 - s^2 \text{ for all } j$$

With the parameter values given in Table 1 both systems fulfil this condition.

Table 2 shows the standard deviation of the log population sizes for each of the populations under the two linkage schemes, estimated from 25 runs of 100 generations each. It may be seen that

(1) The standard deviation is approximately constant for any trophic layer within either linkage scheme.

(2) The level of the standard deviation varies between trophic layers within the linkage schemes.

(3) For all populations the standard deviation is greater in the 'simple' linkage scheme, so that the complex scheme is more stable.

DISCUSSION

The results from the computer simulation indicate less variability and so greater stability in the populations when linked by the complex scheme. This is brought about by the predation and carrying capacity functions being split over a greater number of populations in the complex scheme, each population having an independently varying survival rate. The differences between trophic layers are unimportant since the stability of each trophic layer is determined in part by its position relative to the constant non-fluctuating food sources utilised by trophic layer 3, i.e. stability tends to decrease upward to layer 1. The relative values of R_x were determined so as to keep the standard deviations of each tropic layer at approximately the same order of magnitude.

The results from the neighbourhood stability analysis indicate that it is only when the reproductive rates and the percentage mortality rates are low (the latter in the region of 20 per cent), that the model is stable. This corresponds to the findings of Maynard Smith and Slatkin (1973) for their predator-prey model, that coexistence of predator and prey is only possible where the equilibrium numbers of prey are not substantially below their carrying capacity. The similarity of the neighbourhood stability properties of the two linkage schemes, at least at the low values of B_x and R_x

Table 2. The standard deviation of the log population levels for simple and complex linkages, estimated from 25 runs of 100 generations

Simple linkage

Trophic layer	Population							
	1	2	3	4	5	6	7	8
1	0•0507	0•0566						
2	0•0620	0•0602	0•0607	0•0628				
3	0•0449	0•0478	0•0446	0•0456	0•0444	0•0452	0•0461	0•0459

Complex linkage

Trophic layer	Population							
	1	2	3	4	5	6	7	8
1	0•0427	0•0433						
2	0•0487	0•0481	0•0481	0•0501				
3	0•0408	0•0382	0•0398	0•0414	0•0417	0•0388	0•0414	0•0418

may be explained, to some extent, by the fact that all the populations in a given trophic layer have identical dynamic properties.

The simulation experiments indicate a tendency for increased trophic complexity to increase stability in conditions of a fluctuating environment, where the linkage systems had similar neighbourhood stability properties. What bearing the results from this ecosystem model have on the question of stability and complexity in real ecosystems depends both on the realism of the assumptions made in constructing the model and the validity of the computer simulation determination of the stability properties. The difference equations were formulated to make all the populations in the system density-dependently limited and reliant upon energy flow up through the trophic web. These two properties were considered essential since one of the necessarily characteristic features of real ecosystems is that populations are ultimately prevented from indefinite increase by resource limitation. It is this feature that prevents the formation of explosive positive feedback loops in the population linkages, which tend to arise when model populations are linked at random (Gardner and Ashby, 1970).

The model is, however, unrealistic in that the linkages are over-simplified and regularised and that the carrying capacity and predation functions are linear, taking no account of known decreases in predation efficiency at high and low prey densities. The environmental variation is taken to affect each population independently (since the covariance of S was taken as zero for all populations). Whether such an assumption is valid for natural populations is difficult to assess, since it is not usually possible to distinguish between fluctuations due to factors acting directly upon the survival of the population itself, as modelled by the variation in S, and fluctuations caused by variation in numbers of other populations (particularly prey species) in the ecosystem. It is unlikely that random factors affect ecologically distinct populations identically, and since the two linkage schemes modelled have similar neighbourhood stability properties, the stability of the complex scheme might be expected to remain higher with increasing degrees of correlation between the random fluctuations until the correlation approaches unity.

MacArthur (1955) argued that 'a large number of paths through each species is necessary to reduce the effects of overpopulation of one species' So one reason for expecting more complex ecosystems to be more stable than simpler ones is that the individual populations may be better buffered against chance fluctuations of other populations in the system. The results obtained from this model agree with this conjecture.

ACKNOWLEDGEMENTS

I am indebted to Dr. J. Beddington for his valuable advice and discussion and to D.J. Thompson and P.J. Young who helped with running the computer programmes.

SUMMARY

A difference equation ecosystem model is outlined based on the predator-prey equations of Leslie and its stability properties examined under two food web linkage schemes of differing complexity. A neighbourhood stability analysis, performed for differing parameter values showed no consistent tendency for either linkage scheme to be more stable. The complex linkage scheme proved to be the more stable in a computer simulation of the model behaviour in a randomly fluctuating environment.

REFERENCES

Elton, C. (1958). *The ecology of invasions by animals and plants.* London: Methuen.

Gardner, M.R. and Ashby, W.R. (1970). Connectance of large dynamic (cybernetic) systems: critical values for stability. *Nature, Lond.,* 228, 784.

Leslie, P.H. (1948). Some further notes on the uses of matrices in population mathematics. *Biometrika,* 35, 213-45.

Leslie, P.H. (1958). A stochastic model for studying the properties of certain biological systems by numerical methods. *Biometrika,* 45, 16-31.

MacArthur, R.H. (1955). Fluctuations of animal populations, and a measure of community stability. *Ecology,* 36, 533-6.

May, R.M. (1971). Stability in model ecosystems. *Proc. Ecol. Soc. Australia,* 6, 18-56.

May, R.M. (1973). *Stability and complexity in model ecosystems.* Princeton: Princeton University Press.

Maynard Smith, J. and Slatkin, M. (1973). The stability of predator prey systems. *Ecology,* 54, 384-91.

Reynoldson, T.B. and Davies, R.W. (1970). Food niche and coexistence in lake-dwelling triclads. *J. Anim. Ecol.*, 39, 599-617.
Watt, K.E.F. (1965). Community stability and the strategy of biological control. *Can. Ent.*, 97, 887-95.
Williamson, M. (1972). *The analysis of biological populations.* London: Edward Arnold.

APPENDIX 1

Matrix values of constants $Q_{x,j\rightarrow k}$ and $P_{x,j\leftarrow i}$ for simple and complex linkages

Simple linkage

$Q_{1,j\rightarrow k}$				
j	k=1	2	3	4
1	·5	·5	0	0
2	0	0	·5	·5

$Q_{2,j\rightarrow k}$								
j	k=1	2	3	4	5	6	7	8
1	·5	·5	0	0	0	0	0	0
2	0	0	·5	·5	0	0	0	0
3	0	0	0	0	·5	·5	0	0
4	0	0	0	0	0	0	·5	·5

$P_{2,j\leftarrow i}$		
j	i=1	2
1	1·0	0
2	1·0	0
3	0	1·0
4	0	1·0

$P_{3,j\leftarrow i}$				
j	i=1	2	3	4
1	1·0	0	0	0
2	1·0	0	0	0
3	0	1·0	0	0
4	0	1·0	0	0
5	0	0	1·0	0
6	0	0	1·0	0
7	0	0	0	1·0
8	0	0	0	1·0

Appendix 1 continued

Complex linkage

$Q_{1,j \to k}$

j	k=1	2	3	4
1	·5	·25	·25	0
2	0	·25	·25	·5

$Q_{2,j \to k}$

j	k=1	2	3	4	5	6	7	8
1	·5	·125	·125	0	0	·125	·125	0
2	0	·125	·125	·5	0	·125	·125	0
3	0	·125	·125	0	·5	·125	·125	0
4	0	·125	·125	0	0	·125	·125	·5

$P_{2,j \leftarrow i}$

j	i=1	2
1	1·0	0
2	·5	·5
3	·5	·5
4	0	1·0

$P_{3,j \leftarrow i}$

j	i=1	2	3	4
1	1·0	0	0	0
2	·25	·25	·25	·25
3	·25	·25	·25	·25
4	0	1·0	0	0
5	0	0	1·0	0
6	·25	·25	·25	·25
7	·25	·25	·25	·25
8	0	0	0	1·0

Part four: Spatial studies between trophic levels

Stability of plankton ecosystems

JOHN STEELE

Marine Laboratory, Aberdeen

The problem of stability of ecosystems is usually dealt with theoretically by assuming the populations to be uniform in space and considering only variations with time. Yet there are schools of thought which maintain that the spatial heterogeneity of the physical environment which produces patterns of patchiness in these populations can be as important a factor in ensuring long-term survival of a species (MacArthur and Wilson, 1967; Andrewartha and Birch, 1954). The theoretical problem concerns the simulation of this variability in the environment. On land, the problem may be intractable but in the sea, on certain scales at least, and in relatively simple situations, the lateral turbulent diffusion of the water is a dominant physical process and can be expressed in a mathematical formulation. Thus the interaction of this process with biological events can be studied.

The great variability in concentration of phyto- and zooplankton is well established. At one extreme it is found in the variance of replicate samples (Cassie, 1963) and on the large scale in differences over the North Sea (Glover *et al.*, 1970), or North Atlantic (Bé *et al.*, 1971). The problem of 'patchiness' appears to arise at intermediate or meso scales in space and time. Cushing and Tungate (1963) followed a *Calanus* patch in the southern North Sea for nine weeks and the patch had dimensions of the order of 100 km. Detailed surveys in the northern North Sea showed similar gradients in both zooplankton and in chlorophyll, an indicator of phytoplankton. These did not appear to be

Ecological Stability
edited by M.B. Usher and M.H. Williamson.

related to variation in such physical parameters as salinity or temperature (Steele, in press b). Such observations raise the question of whether patches arise from physical or biological causes.

The main question in the construction of theoretical pictures of plankton dynamics concerns the complexity of ecosystem structure used to represent these populations and the type of behavioural responses required. Any natural system contains many species at each trophic level. Further, the feeding response of any particular herbivorous zooplankton to each of a range of phytoplankton species may be different and each response may be a non-linear function of the density of that species. Some simplification is needed and there are really two choices: a large number of species with simple interactions or a simple food chain with complex relations between each level. The latter appears more useful when considering stability problems (Steele, in press b) and is developed here in relation to the added effects of diffusion. In this context 'stability' is taken to mean the ability of a system to return to a steady state or a limit cycle after some perturbation. In a natural system the perturbations are usually thought to be caused by fluctuations in the physical factors which determine rates of phytoplankton growth. Such factors are, for example, sunlight, mixed layer depth or turbidity of the water. An example of a major 'perturbation' is the establishment at the end of winter of a shallow thermocline in temperate waters which starts the spring outburst and is followed by a period of less variation in the summer. Such a major change in conditions can be simulated by a model which is designed for running on a computer where a large number of details of phytoplankton growth, zooplankton reproduction and predation can be included, and which generates a limit cycle (Steele, 1972). For a mathematical analysis, much simpler interrelations have to be used and the perturbations must be very small in relation to the steady state values.

PHYTOPLANKTON-ZOOPLANKTON INTERACTIONS; LINEAR THEORY

The aim of the equations developed here is to show the relative effects of functional responses and of diffusion on stability. Thus the formal relations are developed to state these explicitly. Consider populations of plants, P, and herbivores, H, varying along one horizontal dimension, x

$$\frac{\partial P}{\partial t} = aP - f_1(P)H + \frac{\partial}{\partial x} k \frac{\partial P}{\partial x} \tag{1}$$

$$\frac{\partial H}{\partial t} = f_2(P)H - f_3(H) + \frac{\partial}{\partial x} k \frac{\partial H}{\partial x} \tag{2}$$

where a is the growth rate of plants per unit density, $f_1(P)$ is the grazing rate of H on P, $f_2(P)$ is the growth rate for herbivores, $f_3(H)$ is the predation rate on the herbivores, assumed to be a function of H only, and k is the lateral eddy diffusivity for plants and herbivores. The major assumption which must be made is that there exists a steady state solution of these equations independent of x

$$aP_0 = f_1(P_0)H_0 \tag{3}$$

$$f_2(P_0)H_0 = f_3(H_0) \tag{4}$$

If we take small perturbations p, h such that

$$P = P_0 + p$$

$$H = H_0 + h$$

then from Equations 1 and 2

$$\frac{\partial p}{\partial t} = \alpha p + \beta h + \frac{\partial}{\partial x} k \frac{\partial p}{\partial x} \tag{5}$$

$$\frac{\partial h}{\partial t} = \gamma p + \delta h + \frac{\partial}{\partial x} k \frac{\partial h}{\partial x} \tag{6}$$

where

$$\alpha = a - f_1'(P_0)H_0 \ , \ \beta = -f_1(P_0)$$

$$\gamma = f_2'(P_0)H_0 \ , \ \delta = f_2(P_0) - f_3'(H_0)$$

From Equations 3 and 4

$$\alpha = [f_1(P_0)/P_0 - f_1'(P_0)]H_0 \tag{7}$$

$$\delta = f_3(H_0)/H_0 - f_3'(H_0) \tag{8}$$

Solutions ignoring diffusion

Consider first perturbations not containing x, i.e. ignoring the effects of diffusion, then the solution is in terms of exponentials whose coefficients are given by the roots of the equation

$$\begin{vmatrix} \alpha - \lambda & \beta \\ \gamma & \delta - \lambda \end{vmatrix} = \lambda^2 - (\alpha + \delta)\lambda + \alpha\delta - \beta\gamma = 0$$

For these roots to be negative, i.e. for the perturbations to be damped out and so for the system to be stable

$$\alpha + \delta < 0 \tag{9}$$

$$\alpha\delta - \beta\gamma > 0 \tag{10}$$

Now we assume

$$f_1(P) > 0, \text{ i.e. } \beta < 0$$

$$f_2(P) \text{ increases as } P \text{ increases, i.e. } \gamma > 0$$

and hence it follows that $\beta\gamma < 0$.

If $\alpha < 0$ and $\delta < 0$ then both Equations 9 and 10 are satisfied. If either is positive then the conditions for stability will depend on the numerical value of the coefficients. Thus we need to examine more closely the biological implications of these coefficients. Both have the same form. The sign depends on the terms of the form

$$F = f(X_0)/X_0 - f'(X_0)$$

where $X_0 = P_0$ or H_0. Graphically there are three cases, as shown in Fig.1 for $F = 0$, $F > 0$, $F < 0$. These can be associated with three general shapes to the curve relating grazing or predation to the density of the food, Fig.2. The first of these is the simple relation $f(X) = X$, i.e. grazing or predation is proportional to the food concentra-

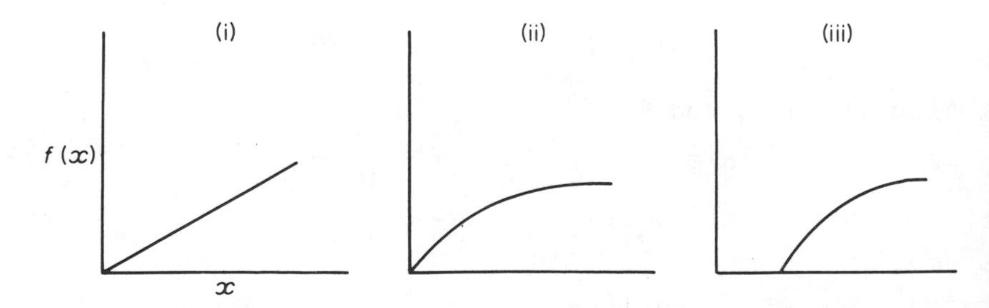

Fig.1. Stability criteria for functional relations of the form $f(x)$. (i) $F = 0$, (ii) $F > 0$, (iii) $F < 0$

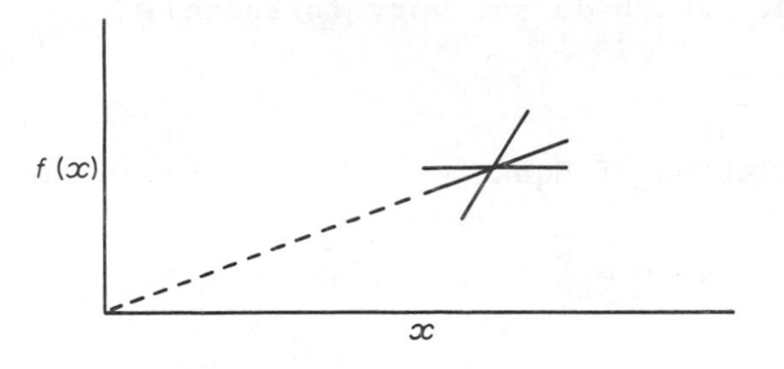

Fig.2. Three general functional responses satisfying the three criteria of Fig.1

tion. This is the simplest assumption which is used in the Lotka-Volterra equations where Equations 1 and 2 would reduce to this form with $f_1(P) \sim P$, $f_2(P) \sim P$, $f_3(H) \sim H$.

It is generally accepted that this is too idealised a relation since in all cases which have been investigated in detail an organism cannot continuously increase its food intake exactly in proportion to the increase in food density. Rather the intake slows down and then levels off, Fig.2, line (ii).

If this process were the sole response of a predator to increasing prey density or a herbivore to increasing plant food, then $F > 0$ for all values of X and for both trophic levels, and the response to perturbation would be unstable, within the terms of the system as defined here. The third possibility is that the response of $f(X)$ to X has the shape shown in Fig.2, line (iii), which can, within a certain range of values of X, give $F < 0$. There is now sufficient evidence to show that this can occur in many cases, and it is accepted that this response helps to stabilise the ecosystem in which it occurs (Holling, 1965). I have proposed (Steele, in press b) that the available evidence suggests that although this type of response is usually found at the herbivore-predator level in terrestrial systems, it is most important at the plant-herbivore level in the sea. This 'threshold' response indicates more complex behaviour patterns and is often associated with selective feeding. If both the herbivore and the predator display this pattern around the steady-state (P_0, H_0) then the system is unconditionally stable and by extrapolation, similar larger food chains, or food webs, may also be stable. The question which arises at this point is whether exchange processes in the physical environment can alter these conclusions reached for a model involving events at one point only (or equivalently, events which do not vary in space).

Solutions with diffusion

The full solution of Equations 5 and 6 is of the form

$$p = \sum_{1}^{\infty} A_n \exp\,(i\lambda nx + \mu_n t) \qquad (11)$$

$$h = \sum_{1}^{\infty} B_n \exp\,(i\lambda nx + \mu_n t) \qquad (12)$$

leading to the condition

$$\mu_n^{\,2} + \mu_n(2\lambda^2 n^2 k - \alpha - \delta) + (\lambda^2 n^2 k - \alpha)(\lambda^2 n^2 k - \delta) - \beta\gamma = 0$$

From this the requirements for positive stability are

$$2\lambda^2 n^2 k > \alpha + \delta$$

$$(\lambda^2 n^2 k - \alpha)(\lambda^2 n^2 k - \delta) - \beta\gamma > 0$$

These reduce to Equations 9 and 10 when $k = 0$. Further for $\alpha < 0$ and $\delta < 0$ then the conditions hold for all values of k and λ. Diffusion becomes important when $\alpha + \delta > 0$. This means that the solution without any diffusion term would be unstable, Equation 9, but with diffusion introduced the system can be stable for certain values of λ and k. The most critical condition is for $n = 1$. Since λ can be expressed as $\lambda = 2\pi/L$ where L is the wavelength, then the critical factor is the largest wavelength required for the Fourier series, Equations 11 and 12, to portray the perturbation. In other words, if the system without turbulence is unstable then for short wavelengths diffusion can stabilise the system but there will be some critical wavelength, L^* above which the system again becomes unstable. The numerical value of this wavelength is very important in understanding on what scale turbulence can act as a controlling factor in ecosystem stability. This was recognised by Kierstead and Slobodkin (1953) who developed a simple form of the solution used here, and applied it to red tides. In this form it was assumed that the phytoplankton causing the red tide were not being grazed, thus there was no equation in H and $f_1(P) = 0$ giving

$$\frac{\partial p}{\partial t} = ap + k\,\frac{\partial^2 p}{\partial x^2}$$

as the perturbation equation, leading to the condition

$$L^* = 2\pi(k/a)^{0 \cdot 5}$$

for the critical value of L in relation to k and a. For the patch to increase in amplitude 'causing' a red tide, $L > L^*$. The numerical problem, then, was to put a value to the diffusion coefficient, k. During the last twenty years there has been considerable experimental work on lateral turbulent diffusion using the spreading of Rhodamine B dye. This work in coastal waters together with the study of certain natural phenomena has provided values for dimensions in the range $10^4 - 10^8$cm. Since the dye release has certain physical similarities to 'patchiness' in plankton, the value obtained should be applicable to the latter problem. The main feature of the results is the fact that the parameter k is not a constant but is dependent on the scale on which the observations are made.

A comparison of the observed values of k (Okubo, 1971) with a range of probable phytoplankton division rates of 0•1-1•0 per day suggests that, for open water, the critical value L^* may lie in the range 3-100 km (Steele, in press c). However, the value of k can depend on many transient features such as wind, thermocline structure, currents (including vertical velocity shear). Also any restrictions imposed by coastal boundaries will tend to reduce the scale of turbulence and so decrease L^*. As a possible general conclusion, one would expect smaller scale fluctuations to be under the control of turbulent effects and Platt (1972) has provided evidence for this. It would be only at scales of the order of tens of kilometres that 'biological' patchiness can occur. (I am neglecting here the special phenomenon of 'wind-rows' produced by Langmuir circulation). For example, the spring outburst can be expected to occur over large areas of, say, the North Sea or northern North Atlantic, and could have division rates of greater than 0•3 per day and would, initially at least, have very low grazing rates by the herbivores. Under these conditions the prediction would be that patches could occur with typical wavelengths in excess of 10 km. Below that any perturbations would be damped out. Obviously these larger patches would tend to increase in amplitude indefinitely unless some other processes were involved. Once the spring outburst is over the levels of plants and herbivores are determined more by other factors. This is shown by Equations 7 and 8 from which

$$\alpha + \delta = [f_1(P_0)/P_0 - f_1'(P_0)]H_0 + f_3(H_0)/H_0 - f_3'(H_0)$$

which is a function of the grazing, f_1, and predation, f_3, rates. As discussed earlier each of the two terms in brackets on the righthand side gives the difference between the slope of the line from the origin through (P_0, f_1) and (H_0, f_3) respectively, and the slope of the curve at that point. It is possible to imagine that these differences are positive but so small that the critical dimension L^* in this case is larger than the particular area in question. This may occur in areas where the plant growth rate is very low (i.e. $a = f_1/P_0$ is very small) and the populations of plants are sparse. Such sub-tropical areas as the Sargasso Sea may display this uniformity. However, it is necessary to return to the observations in temperate waters which display patchiness with dimensions of the order of 10-100 km occurring after the spring outburst. It appears that theoretically a relatively simple combination of growth processes and turbulence must lead either to a relatively uniform environment or to instability on some larger scale. The question is whether any smaller scale behavioural patterns could initiate patchiness at intermediate dimensions.

PHYTOPLANKTON-ZOOPLANKTON INTERACTIONS; NON-LINEAR THEORY

One problem arises from the mathematical form of the analys is used so far. The equations are linearised and used to study the effects of very small perturbations. Thus the effects of any perturbation are found only at the wavelengt of that perturbation. However, the essential feature of an equations involving biological interactions is that they ar non-linear, i.e. they involve terms such as $P \times H$ rather than merely $P + H$, etc. Further, perturbations such as the spring outburst are large compared with the steady state. It is inevitable that these two facts will have consequence not included in the linear perturbation analysis used so far. To give some indication of these consequences conside the simple pair of equations; the Lotka-Volterra relations with diffusion added and a grazing threshold P^*

$$\frac{\partial P}{\partial t} = aP - b(P-P^*)H + \frac{\partial}{\partial x} k \frac{\partial P}{\partial x} \qquad (13$$

$$\frac{\partial H}{\partial t} = c(P-P^*)H - dH + \frac{\partial}{\partial x} k \frac{\partial H}{\partial x} \qquad (14)$$

This form of grazing, as indicated, is unrealistic at higher plant concentrations; also it does not hold for $P < P^*$. However, in the neighbourhood of P_0 it gives a first approximation to the conditions given in Fig.2, including the unstable condition (ii) by taking $P^* < 0$. The simple relation without P^* has already been described (Steele, in press a). Assume reflexion at the boundaries $(0,L)$ and consider a particular perturbation about a steady state $P = A_0$, $H = B_0$, then

$$P = \sum_0^\infty A_n \cos \lambda n x$$

$$H = \sum_0^\infty B_n \cos \lambda n x$$

$$\lambda = 2\pi/L$$

$$\frac{dA_n}{dt} = A_n(a - \lambda^2 n^2 k) + bP^*B_n - bC_n$$

$$\frac{dB_n}{dt} = cC_n - B_n(d - cP^* + \lambda^2 n^2 k)$$

and since $\cos \lambda r x \cos \lambda s x = \frac{1}{2}\{\cos \lambda(r+s)x + \cos \lambda(r-s)x\}$

$$C_n = 0{\cdot}5\left[\sum_{r+s=n} A_r B_s + \sum_{|r-s|=n} A_r B_s\right]$$

As has been pointed out (Steele, in press a) this means that short wavelength perturbations can have effects at higher wavelengths and thus can 'pass through' the critical wavelength. In particular the equations for A_0, B_0 are

$$\frac{dA_0}{dt} = aA_0 - b(A_0 - P^*)B_0 - 0{\cdot}5\, b \sum_{r=1}^{\infty} A_r B_r \qquad (15)$$

$$\frac{dB_0}{dt} = c(A_0 - P^*)B_0 - dB_0 + 0{\cdot}5\, c \sum_{r=1}^{\infty} A_r B_r \qquad (16)$$

Here A_0 and B_0 represent the average of P and H throughout the interval $(0,L)$ and thus Equations 15 and 16 are the equations for the mean conditions but with an added term dependent on the perturbations. It is this addition of the spatial variation which is usually ignored when 'average' conditions over some area are used, or even when only the linearised equations involving diffusion are considered.

The term $\sum A_r B_r$ has a direct interpretation in terms of the fluctuations in P and H since the correlation coefficient for P and H, given by Davis (1941), is

$$\sum A_r B_r / (\sum A_r^2 \sum B_r^2)^{0 \cdot 5}$$

Thus, if the perturbations in P and H are to interfere with average conditions, P and H must be correlated and the magnitude of the correlation as well as the amplitude of the fluctuations will determine the degree of interference.

As an indication of the consequences suppose that random fluctuations in the environment impose perturbations on P and H which are correlated and the correlation is assumed to be constant giving a constant coefficient $C_0 = 0 \cdot 5 \sum A_r B_r$. Then it can be shown by the usual perturbation analysis that:

(1) For sufficiently large positive C_0 there is no simple steady state solution, and

(2) For sufficiently large negative C_0, and certain conditions on the other coefficients, the steady state will be unstable (see Appendix).

In both cases the existence of a sufficiently large threshold P^* can reverse the effect. As usual, such a perturbation analysis does no more than indicate certain possible types of effect. However, it is sufficient to show that, in principle, short wavelength fluctuations in P and H if they are of sufficient amplitude and are highly correlated, would produce instabilities. In turn this implies that in any 'world' where there are small-scale fluctuations at several trophic levels each, on their own, would be damped out, but in combination they could cause instabilities in the whole large-scale population structure unless these populations have functional responses to cope with such perturbations. Especially, in looking for the causes of patchiness at a certain scale, we need to consider not only events at that scale but also the possible effects of fluctuations at much smaller scales.

Further, in considering possible types of perturbations, we tend to think of those events in the physical environment which are imposed from outside the system and which may occur at any frequency in space or time. However, the biological cycles may impose changes within the system which can be considered as perturbations. The idea of phytoplankton as a single parameter, P, may have some validity but the herbivores cannot, realistically, be treated in this way. Any individual copepod starts feeding as a nauplius, grows through copepodite stages to an adult which, if female, produces eggs to start the next generation. I have shown that this process, when considered without diffusion, produces temporal cycles in P and the biomass of herbivore H (Steele, 1972). The problem with this work was that, although these cycles fitted, albeit very roughly, to observations in the North Sea, the observations were of spatial distribution rather than temporal changes. These biological cycles may act as small-scale perturbations at the herbivore level and, as a result of differing food requirements, produce variations in the phytoplankton. This aspect will be developed elsewhere.

CONCLUSIONS

From these considerations there are certain general but tentative conclusions about the relation between lateral turbulent diffusion and the stability of planktonic ecosystems:

(1) If an ecosystem is basically unstable when considered without diffusion processes then diffusion can remove the instability at smaller scales but not at larger scales. For moderate to rapid growth of the phytoplankton the critical scale in an open sea environment is of the order of 10-100 km.

(2) The effect of non-linearity in the basic equations is to redistribute perturbations up and down the scale of wavelengths so that small-scale perturbations can affect the whole population levels.

(3) The general requirements for 'threshold' type responses in the grazing of herbivores on phytoplankton, derived for a system without diffusion, will still apply.

(4) Small-scale 'behaviour' of copepods, including their growth cycles, could act as perturbations to generate patches at larger scales.

(5) The critical factor is the correlation between spatial population variation at different trophic levels. The spectral distribution of variance as a function of scale can provide an indication of the scales of patchiness and of the possible effects on stability. Such analyses will pro-

vide tests of the concepts put forward here.

REFERENCES

Andrewartha, H.G. and Birch, L.C. (1954). *The distribution and abundance of animals*. Chicago: University of Chicago Press.

Bé, A.W.H., Forns, J.M. and Roels, O.A. (1971). Plankton abundance in the North Atlantic Ocean. *Fertility of the sea* (Ed. by J.D. Costlow). New York, London, Paris: Gordon and Breach Science Publishers.

Cassie, R.M. (1963). Microdistribution of plankton. *Oceanogr. Mar. Biol. Ann. Rev.*, 1, 223-52.

Cushing, D.H. and Tungate, D.S. (1963). Studies on a *Calanus* patch I. The identification of a *Calanus* patch. *J. mar. biol. Ass. U.K.*, 43, 327-37.

Davis, H.T. (1941). *The analysis of economic time series*. Bloomington: Principia Press Inc.

Glover, R.S., Robinson, G.A. and Colebrook, J.M. (1970). Plankton in the North Atlantic - an example of the problems of analysing variability in the environment. *F.A.O. Tech. Conf. on Mar. Poll.*, FIR:MP/70/E-55.

Holling, C.S. (1965). The functional response of predators to prey density and its role in mimicry and population regulation. *Mem. ent. Soc. Can.*, 45.

Kierstead, H. and Slobodkin, L.B. (1953). The size of water masses containing plankton blooms. *J. Mar. Res.*, 12, 141-7.

MacArthur, R.H. and Wilson, E.O. (1967). *The Theory of island biogeography*. Princeton: Princeton University Press.

Okubo, A. (1971). Oceanic diffusion diagrams. *Deep Sea Res.*, 18, 789-802.

Platt, T. (1972). Local phytoplankton abundance and turbulence. *Deep Sea Res.*, 19, 183-8.

Steele, J.H. (1972). Factors controlling marine ecosystems. *The Changing Chemistry of the Oceans*. Nobel Symposium 20. (Ed. by D. Dyrssen and D. Jagner) pp.209-21. New York, London, Sydney: Wiley.

Steele, J.H. (In press a). Spatial heterogeneity and population stability *Nature, Lond.*

Steele, J.H. (In press b). *The structure of marine ecosystems*. Cambridge: Harvard University Press.

Steele, J.H. (In press c). The state of the art in biological modelling. *Modelling of Marine Systems*, Ed. by Nihoul. Elsevier Oceanography Series.

APPENDIX

It can be shown that the steady state values of Equations 14 and 15 are

$$A_0 = \frac{ad^* \pm (a^2d^{*2} - 4\ abcdC_0)^{0\cdot5}}{2\ ac}$$

$$B_0 = \frac{ad^* \pm (a^2d^{*2} - 4\ abcdC_0)^{0\cdot5}}{2\ bd}$$

where $d^* = d + cP^*$ and the stability conditions are

$$C_0\ (a/d - 1) - P^*B_0 < 0$$

$$A_0B_0 - C_0 > 0$$

The latter holds unconditionally but the former depends on the sign of C_0 and $(a/d - 1)$. The most interesting condition is $C_0 < 0$. Then if $a < d$, since $B_0 = 0(C_0^{0\cdot5})$, there will be a sufficiently large value of C_0 to make the system unstable. Since, biologically, $b > c$ and $acA_0 = bdB_0$, the condition $a < d$ is equivalent to $A_0 > B_0$. That is, the steady state plant population must be greater than the steady state herbivores. This is an acceptable condition.

Index